FOUNDATIONS & CONCRETE WORK

THE BEST OF
Fine Homebuilding

FOUNDATIONS & CONCRETE WORK

THE BEST OF
Fine Homebuilding

The Taunton Press

Cover photo: Bob Syvanen

**Back cover photos: Bob Syvanen (top and center),
Roe Osborn (bottom)**

The Taunton Press
Inspiration for hands-on living®

©1998 by The Taunton Press, Inc.
All rights reserved.

Printed in the United States of America
10 9 8

Taunton's For Pros / By Pros®: Foundations & Concrete Work (The Best of Fine Homebuilding) was originally published in 1997 by The Taunton Press, Inc.

For Pros / By Pros® is a trademark of The Taunton Press, Inc., registered in the U.S. Patent and Trademark Office.

The Taunton Press, Inc.
63 South Main Street
P.O. Box 5506
Newtown, Connecticut 06470-5506
e-mail: tp@taunton.com

Library of Congress Cataloging-in-Publication Data

For Pros / By Pros®: Foundations & concrete work (the best of Fine homebuilding).
     p. cm.
    Includes index.
    ISBN 1-56158-330-8
    1. Foundations.   2. Concrete construction.   I. Taunton Press.
II. Fine homebuilding.
TH2101.F635   1998
624.1'5 — dc20
                                 96-41126
                                       CIP

CONTENTS

(continued)

INTRODUCTION

S TORIES ABOUT BUILDING, like most stories I suppose, are prone to embellishment in their retelling. So the following story may not have happened just this way, but...

Some years ago, a major supermarket chain was building a new distribution center out in Richmond, California. In the midst of pouring a tall concrete wall, the formwork started to give way. Disaster was averted when somebody enlisted the supermarket's fleet of forklifts to brace the wall. The forklifts (one dozen? two dozen?) were distributed along the wall, their prongs raised, bearing against the forms and vibrating like tuning forks as the wet, heavy concrete flowed in.

Well, in case you don't have a fleet of forklifts standing by to brace your concrete forms, you'll find some alternative solutions in this book. Assembled here are more than two dozen articles from past issues of *Fine Homebuilding* magazine. The articles cover everything from footings and foundation drains to pouring concrete walls and slabs to laying up concrete block or even nailing together a wooden foundation. And while none of the authors worked on that supermarket distribution center, they are the same kind of clever and resourceful people as the person who thought to brace sagging formwork with forklifts.

–Kevin Ireton, editor

Concrete

Understanding the characteristics of this material can take some of the anxious moments out of your pour and ensure a finished product of high quality

by Trey Loy

Concrete is a remarkable material that can be cast into almost any shape. It will sustain and transmit tremendous loads, and once hardened, it is practically indestructible. Yet few builders feel as affectionate about concrete as did Slim Gaillard and Lee Ricks in their scat tune from the 1940s.

> Cement mixer, put-ti, put-ti,
> Cement mixer, put-ti, put-ti,
> Puddle-o-votty, puddle-o-goody,
> Puddle-o-scooty, puddle-o-vett.
> Who wants a bucket of cement?
>
> First you get some gravel,
> Pour it in the vout.
> To mix a mess of mortar
> You add cement and grout.
> See the mellow rooney come out.
>
> Slurp, slurp, slurp.
>
> Cement mixer, put-ti, put-ti,
> Cement mixer, put-ti, put-ti,
> I can never get enough
> of that wonderful stuff.

Cement mixers have been largely replaced by huge batch plants that measure out hundreds of cubic yards of ready-mix a day to waiting transit mixers. The leisurely pace of pouring concrete is also a thing of the past. The distant rumblings of an approaching concrete truck can strike fear into the heart of a carpenter still bracing the forms. At nearly two tons a cubic yard, concrete has to be poured immediately, with no time to ponder problems or locate tools. For many people, a pour is considered successful when forms don't break; and a sigh of relief can be heard when a slab is smooth and unblemished by cracks the next day.

Contrary to its reputation, concrete reacts predictably, and the builder can regulate many of the variables that affect its working properties while plastic, and its strength when hardened. Some practical knowledge of the kinds of cements and aggregates and the correct proportions of each, admixtures, slump, and the effects of weather during pouring and curing can make the difference between feeling confident about a pour, and feeling you are constantly dodging disaster.

Concrete is a mixture of water, portland cement, and fine and coarse aggregates. The active ingredients are water and cement. They combine in a chemical reaction called hydra-

The advent of ready-mix has brought a change to the quality, quantity and pace of concrete pours. The physical properties of concrete in its plastic and hardened states are well understood by researchers, engineers and batch-plant operators, but this information seldom trickles down to the builder who is actually working with the material. Knowing how mixes are designed, and the different kinds of cements, aggregates and admixtures that are used to alter the concrete can give the builder the ability to predict how it is going to react and why. For more on hand- and machine-mixing concrete on site, see FHB #13, pp. 34-35.

tion. Although concrete that is beginning to set appears to be drying out, or dehydrating, about half the water is actually incorporated in the hydration process and becomes a permanent part of the bonding paste. This is why concrete needs to be kept moist during the first few days of curing. In fact, concrete will harden quite effectively under water.

Aggregates—The major function of aggregates is to make concrete more economical. While the cement paste binds the aggregates together in a solid mass, the aggregates keep the concrete from shrinking and cracking as hydration and evaporation take place during setting and hardening. Neat, or pure, cement is not nearly as strong as concrete correctly proportioned with aggregate.

Fine aggregate can be either sand or rock screenings. Fine-aggregate particles range in size from very fine sand to ¼ in. Coarse aggregate is either gravel or crushed stone ¼ in. to 1½ in. in diameter. The aggregate mix should be proportioned so that the smaller particles fill in the voids between the larger ones. For thick foundations and footings, gravel or rock with a diameter of 1½ inches is used. For ordinary walls, the largest pieces should be not more than one-fifth the thickness of the finished wall section. And for slabs, the maximum thickness of the rock aggregate should not be more than one-third that of the slab.

The amounts and sizes of the sand and gravel are also adjusted for the strength and workability of the mix. The plasticity of concrete in its wet state, as well as its ultimate density, depend in part on the aggregates meshing. How well the mud moves down the chute, how easily it fills the forms, and how well it finishes are measures of its workability.

Cement—An English stonemason, Joseph Aspdin, patented the process for manufacturing portland cement in 1824. He used this name because it produced a hardened concrete with a color that reminded him of the natural grey stone on the Isle of Portland. Portland cement was first produced in North America in 1872. Interestingly, the use of steel reinforcing bar was introduced a few years later, around 1880. And the first patent for prestressed concrete using steel-wire rope was issued in 1888.

Today, portland cement is produced in huge rotary kilns at temperatures of nearly

2,700°F. At this temperature, lime, silica, alumina and iron oxides, which are derived from limestone, oyster shells, marl, shale, iron ore and clay, undergo a kind of molecular reformation called calcination. After cooling, the resulting greenish-black clinkers are pulverized with small amounts of gypsum to control the set time of the cement.

The American Society for Testing and Materials (ASTM) recognizes five types of portland cement. Each is intended for a specific purpose, although they all achieve about the same strength after curing for three months.

Type I. This is the most common type of general-purpose cement, and is used when a specific type isn't called out. Most residential construction uses Type I.

Type II. A moderately sulfate-resistant cement, it is sometimes specified for walkways where de-icing chemicals will be heavily used. It sets more slowly than Type I, an advantage during the summer, when getting a finish on concrete can be a real race. Also, because it generates less heat during curing than Type I, it is better suited for mass pours, where heat radiating from hundreds of yards of curing concrete can cause problems.

Type III. This type is called high-early-strength cement because it achieves most of its strength within the first week of curing. This is useful if the concrete has to be put under full load within a few weeks of pouring, or if forms have to be stripped early. It is not widely stocked by concrete companies, but adding an extra bag of Type I or Type II cement per cubic yard of concrete and mixing at high speeds will produce similar results.

Type IV. A slow-curing variety that generates very little heat by hydration, it is used exclusively in mass concrete, such as dams.

Type V. This cement will withstand severe sulfate action that occurs in heavily alkaline soil or groundwater. Concrete can deteriorate because of physical and chemical reactions between sulfates and compounds formed by hydrated portland cement. Type V gains strength much more slowly than Type I.

The Canadian Standards Association (CSA) has three categories—normal, high-early-strength, and sulfate-resisting. These correspond to ASTM Types I, III, and V.

There are a number of ways that these cement types can be altered to meet special conditions. Portland cement is normally grey. White portland cement, light in color because it's made with a minimum of iron and magnesium oxides, can be tinted by adding pigments or used as is. Blast-furnace slag and pozzolan are two materials that are ground up and blended with portland cement to bring down its cost. When pozzolan is added to Type I it is designated IP; for slag, the abbreviation is IS. Another application of portland cement is

Jitterbugging a slab settles the large aggregate just below the surface to allow a smooth, troweled finish. Most problems in residential concrete are surface faults—crazing, dusting, scaling, honeycombing and shrinkage cracking—rather than strength failures.

lightweight concrete, which uses artificial aggregates and gas-forming admixtures to make it lighter. Types I, II, and III are available with air-entrainers (discussed under admixtures) interground, and designated with an A after the type number.

Portland cement is usually packaged in paper bags. Each bag weighs 94 lb. and contains 1 cu. ft. of cement. A common unit of measure in the past was the barrel, which contained the equivalent of four bags.

The cement content of a mix has a lot to do with its strength. One method of ordering ready-mix concrete is to specify how much cement should be used for each cubic yard of concrete. Producers of ready-mix prefer that you give cement content by weight (such as 470 lb. per cu. yd.), but ordering a certain number of bags, or sacks, of cement per cubic yard (such as a five-bag mix) is still very common. A four-bag mix is the minimum for most residential uses; five-bag is better. You should ask for six-bag or seven-bag if you use smaller aggregate or if you want greater strength, waterproofing and durability.

Water—For mixing concrete, it's best to use water that is fit to drink. It should not contain any oil, alkali or acid. In hydration, the water and cement in concrete combine chemically to form a paste that binds the aggregates together. It can be thought of as a glue.

The more water added to a given amount of cement, the weaker the concrete will ultimately be. This relationship of water and cement is known to concrete engineers as Abrams' law, or the water/cement ratio (W/C). It is a central factor in the design of the mix. The W/C ratio needs to be adjusted for a large number of variables, including the quantity of water that the aggregates are carrying, and the ambient temperature and humidity at the time of the pour.

The strength of concrete decreases as the W/C ratio increases, as seen in the chart above. The first column expresses the ratio in

weight (pounds of water divided by pounds of cement). The second column gives the same ratio in gallons of water per bag of cement.

W/C (weight)	Gallons per bag	Approx. 28-day strength (psi)
.45	5.0	5,000
.49	5.5	4,500
.53	6.0	4,000
.57	6.5	3,500
.62	7.0	3,000

Strength and quality—The strength of a given sample of concrete is measured by how much compression-loading a test cylinder of concrete 6 in. in diameter by 12 in. high can take before fracturing. The testing is done in a laboratory using a hydraulic piston hooked up to a meter that measures pounds per square inch (psi). Compression-strength figures for concrete refer to tests conducted after 28 days of curing unless otherwise noted. These figures indicate how well a batch of concrete will stand up to vertical and lateral loading. It is also a measure of durability and watertightness. Most engineers require that any load-bearing concrete achieve a minimum 28-day strength of 2,500 psi.

Although most concrete is batched to exact engineering standards at the plant, a lot can happen to affect its quality before, during and after the pour. How much water is added to the concrete after it's initially mixed is a good example of this.

Some drivers will ask if you want the mix stiff or sloppy, and then judge how much water to add, according to the slope (if any) involved in the pour, the angle of the chute from the truck to the forms, the depth of the forms, the amount of rebar, how long the pour is taking, and the weather (hot, dry days call for more water). A thin, watery mix is easier to handle, but much of the water added to the concrete at the job site may not have been figured in the mix design, and the resulting concrete will be less durable and much more subject to cracking.

Slump—To regulate the amount of water in ready-mix, specify slump. If you are working from a set of engineered plans, this may already be listed in the specs. Slump is a measure of the consistency of concrete—the higher the slump, the wetter the mixture. Slump is measured with a 12-in. high truncated metal cone with a base diameter of 8 in. and a top diameter of 4 in. The cone is filled with concrete right off the truck and rodded with a tamping rod. Then the cone is lifted free, inverted and placed beside the sagging pile of concrete for comparison. The distance the

Concrete is one of the few residential building materials that have no forgiveness, and any pour will have its frantic moments. But good forming, careful planning, ordering the right mix, and knowing how the mud will react can keep the panic to a minimum.

concrete subsides from the top of the cone, measured to the quarter inch, is the slump.

Roadways, industrial floors and any concrete that is consolidated with mechanical vibration requires a 1-in. to 3-in. slump. Most foundations, slabs and walls consolidated by hand methods such as spading can have a slump between 4 in. and 6 in. On most residential pours, taking the time for a slump test isn't feasible. I usually just eyeball the mix. A good 3-in.to 4-in. slump mix will stand in a pile. A 5-in. to 6-in. slump sags into a blob, and a 7-in. to 8-in. slump just flattens out.

Many problems in residential concrete are the result of high slump. Unlike public projects such as bridges and highways, where compressive strength is essential, residential pours seldom suffer failures under a load. Instead, it is surface faults—honeycombing, crazing, dusting, surface cracking and scaling—that cause the problems.

One serious failure of concrete associated with high slump is segregation, which is a re-separation of concrete back into sand and gravel. Bleeding is another kind of separation that can be serious if it occurs on a large scale. Bleeding is the emergence of water on the surface of newly poured concrete. This occurs when the large aggregate settles within the mass, displacing the water in the mix. Heavy bleeding greatly dilutes the cement particles on the surface of the concrete, making it susceptible to abrasion. When this bleed water is troweled into the surface of a slab during finishing, it can result in crazing lines, shallow parallel fissures called plastic shrinkage cracking, a powdery dusting on the surface, and even scaling, which is the flaking of the finished concrete. If you keep slump low, and finish and cure the resulting concrete with care, you can avoid these problems.

The mix—Although all concrete consists of the same basic ingredients, how they are proportioned can make a huge difference in strength and workability. Because of all the variables involved, there are hundreds of possible combinations that will produce concrete with a wide variety of characteristics. As explained on p. 13, it's very important to mention the conditions and requirements of a pour when ordering concrete, so that a mix can be designed or chosen from standard designs that will give you the kind of concrete you need.

The design of a mix is complex because of the interrelationships of the materials. For example, if the size of coarse aggregate is limited in order to pour a thin slab, this will affect the amount of cement needed to reach the necessary compressive strength. The amount of cement used in turn affects the amount of water to be added, as well as the

Mechanical vibrators should be used with caution, and low-slump concrete. Prolonged vibrating can cause the concrete to segregate, bringing the fines—cement and water—to the top, leaving the heavier aggregates below. This can lead to surface failures on a slab.

Photo: Bob Syvanen

size and proportion of fine aggregate. The mix will also have to be adjusted for slump, workability, admixtures and the weather.

Admixtures—There are four kinds of admixtures that can give concrete specific qualities. The first, called air-entraining agents, are known to most builders who work in areas with hard freezes. This admixture is a material that stabilizes bubbles formed by air incorporated in the concrete during the mixing process. The bubbles create tiny voids that act as expansion chambers or shock absorbers, which allow the concrete to withstand freeze-thaw cycles. The amount of air in the mix is a variable of mix design. It is typically 5% to 7% by volume.

Although these air bubbles make the mix slightly weaker, they also have beneficial effects. Air-entrainers increase the workability of the mud (so less water can be used for a given slump), make the mud more resistant to salts, and produce a more durable concrete.

A set-retarder may be added to ready-mix to prolong its setting time by 30% to 60%. If you need extra working time on hot days, tell the dispatcher when you order, and the plant engineer will determine the exact amount according to the weather and the mix.

Concrete companies are also likely to add a water-reducing agent, sometimes called a plasticizer, which may allow as much as a 15% reduction in water content for a given slump. Water-reducing agents can help minimize problems relating to an excess of water, such as segregation, plastic shrinkage cracking, crazing and dusting. It can also increase the concrete's strength and its bond to steel reinforcing rod.

Probably the best-known admixture is calcium chloride. This chemical is an accelerator, used to get an early set in freezing weather. Ideally, concrete should be poured when it's at least 50°F, with the mud maintained in the forms at 70°F. But pours in cold weather are often necessary, and quite common. If concrete freezes while it's setting or during the first few days of curing, it won't gain much strength and problems will develop. Pop-out, scaling and cracking occur when water in the concrete freezes and expands nearly 9% of its liquid volume.

Contrary to myth, accelerators aren't effective as antifreeze. Concrete, even with calcium chloride added, can freeze. Like any accelerator, it will only decrease the setting time. This allows the builder to pour, finish and insulate the concrete before the onset of freezing temperatures. Calcium chloride should be used sparingly and not just for convenience when better scheduling would solve the problem. It attacks aluminum conduit, lowers the resistance of the concrete to sulfates, increases shrinkage, and generally weakens the mix. If you need to use it to beat the weather, limit it to 2% by weight of cement. An effective alternative is to add an extra bag of cement to each cubic yard of concrete.

Working concrete—If you pour a lot of concrete, buy a pair of rubber boots. Leather work boots will rot off your feet after a few dunkings in concrete. The only way I've found to restore the flexibility of the leather is to remove all of the concrete and soak the boots in motor oil for a day—crude but effective.

Gloves are another must. The cement contains lime, an alkali that dries out skin and leaves it cracked and sore. Thick rubber gloves offer good protection, but they are awkward to wear. Lately I've been wearing doctor's disposable examination gloves. They cost less than $10 for fifty pairs, and fit like another layer of skin.

It's a good idea to keep a bottle of vinegar with your concrete finishing tools. Vinegar contains acetic acid, which neutralizes the lime. Wash your hands in it when you quit for the day, and don't rinse it off immediately. You'll smell like a salad bar, but your hands won't be any the worse for wear the next day.

There are several general procedures that should be followed to end up with strong concrete that looks good after the forms are stripped. Concrete should be poured in horizontal layers of 6 in. to 12 in., depending on the stiffness of the mix. Start in the corners of the forms and work toward the middle. Concrete should not be dumped into separate piles and then leveled and worked together. Pouring it near its final place will save your back and keep the mud from separating.

If you are ordering ready-mix concrete,

check the approaches that the transit mixer can make, and calculate how many chute sections you will need. Most trucks carry 10 ft. of chute. You can usually arrange with the dispatcher to have the driver bring another 10 ft. If this doesn't do it, you will have to hire a pumper, build a chute, or truck the wet mud around the site in wheelbarrows. These alternatives are listed in order of preference. Although everybody has had to use a wheelbarrow to complete a pour on occasion, it is slow, risky work.

Most concrete companies will allow between 30 minutes and an hour to empty a full truck (about 8 cu. yd.), before they charge overtime. Money isn't the only issue. Time is also a critical factor. Depending on the air temperature, the mix can agitate in the drum of the truck for up to an hour after batching. After that, more water has to be added, which will weaken the mix. Under average conditions, concrete that is left in a truck for longer than 90 minutes is considered unusable.

Pumpers should be considered for a job where ready-mix is used and some part of the pour is inaccessible by ordinary means. Hillsides, high walls, muddy ground where a fully loaded truck could become stuck, and sites with dense trees or landscaping are all good candidates. In addition to the huge pump trucks with articulating booms or snorkels used for big commercial jobs, there are smaller portable pumping machines and trucks that are suitable for residential jobs. In most areas, they can be hired with an operator for $100 to $200 for an average pour.

Using a pumper can cut down the number of people needed to make the pour, and still give better results. Small pumpers use a 3-in. or 4-in. diameter hose. This requires using pea gravel as large aggregate and adding an extra bag of cement for each yard. This mix is rich and easy to work. If you need to build a chute for the site, use at least ¾-in. plywood and lots of bracing. Pitch it at about a 5-in-12.

Once the mud is in the forms, it needs to be tamped or vibrated to eliminate voids around rebar or against the form faces. Spading the sides of the form just after the mix is placed will minimize honeycombing and sand streaking, and keep aggregates from showing on the surface of a poured wall. You can buy special spading tools—thin, flat pieces of metal mounted on long handles—but 1x4s or long pieces of rebar work fine. Another good technique to eliminate honeycombing is to rap the forms sharply with your hammer, moving up and down the forms. This brings the cement paste out to the surface of the wall to cover the aggregate.

The best tool for settling the concrete into forms is a vibrator, a portable motor with a long, flexible, waterproof shaft. You can rent one for less than $20 a day. Use a vibrator as you are pouring on each level, particularly at the face of walls, corners and around rebar. Plunge the shaft into the mud every few feet along the length of the form, and let it vibrate for 5 to 15 seconds. Prolonged vibrating is not good because it can cause segregation. In fact, a vibrator shouldn't be used on any mix that can be placed and consolidated readily by hand tools.

With a slab, it is common practice to tamp the wet concrete with a Jitterbug®—a tubular-steel frame with a mesh bottom and waist-high handles—to settle the aggregate below the surface (photo, previous page). This aids in floating and finish troweling, and brings excess water to the surface to evaporate.

Curing—Proper curing is essential in achieving high-strength concrete and durable surfaces. As long as it is kept moist and warm—above 80% relative humidity and 70°F is ideal—concrete will harden indefinitely at a diminishing rate, as shown in the chart below. If the humidity drops below 80%, the surface of the concrete begins to lose moisture more rapidly than the interior of the pour. This causes the surface to shrink, and results in a soft, dusty skin that is less resistant to abrasion. Surface hairline fissures (crazing) are sure signs that concrete dried out too much during its initial curing. Plastic shrinkage cracks, which seldom have any structural significance but mar the appearance of a finished slab, are also caused by allowing the surface of the concrete to dry out. To get the best cure, the surfaces of the concrete should be kept moist for at least three days. After seven days it will have attained about 60% of its eventual 28-day strength.

Concrete can be kept wet by spraying it lightly with water several times a day; or, in the case of a slab, by maintaining a pond of water on the finished surface. How much moisture it will need depends upon air temperature and humidity. It's impossible to keep concrete too wet during curing. Spreading burlap or straw on the surface and soaking it with water will help hold moisture, as will covering the surface with plastic sheeting.

Membrane curing compounds, which can be sprayed on the surface, are also available. Clear compounds are preferred for surfaces that will be exposed. Black curing compounds have an asphaltic base and are used when staining isn't important. The black compounds will also hold in heat as the concrete cures. White compounds are effective in hot weather to reflect heat from the sun.

In cold weather, concrete slabs should be protected against freezing with straw covered with plastic sheeting or insulating blankets. If temperatures are in the 40°s, maintain this insulation for at least 48 hours. Insulated forms used for columns and walls should remain in place for at least a week. If temperatures are lower, keep the covering on even longer. Since hydration is an exothermic reaction, the primary concern should be holding this heat in. Only if the air temperature drops below 0°F should you supply heat. □

Trey Loy is a designer and builder who lives in Little River, Calif. He spent five years pouring concrete professionally.

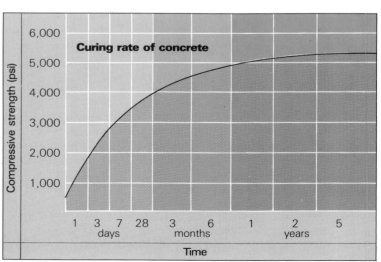

Curing. Concrete acquires most of its strength in the first month after the pour. This graph plots the approximate compressive strength of a six-bag mix using five gallons of water per bag over time, assuming the concrete is maintained at about 70°F.

Photo: Ross Lowell

Figuring and ordering ready-mix concrete

Gulping is the initial reflex for most builders, seasoned or not, when the dispatcher asks the inevitable question, "How many yards you want?" Like it or not, you've got to give an exact figure in cubic yards, no matter how complicated or irregular the pour is. Ordering too much means the driver of the transit mixer will need a place on your job site to dump the excess, which, like the mud inside your forms, will cost you nearly $50 a cubic yard. Ordering too little is even worse. Unless the plant can send another truck with a short load right away (this costs extra), you'll have to cap off your forms, add a keyway, and create a construction joint in a pour that was designed to be monolithic.

The way to prevent these problems is to figure your needs precisely, and then cheat the moment of truth as much as you can. If you have to reserve the concrete several days ahead, give the dispatcher a figure 15% higher than what you think you'll need, and tell him you'll call in the exact number of yards an hour or so before the pour. This way you'll be calculating trench depths and form widths as they exist. If you are figuring a big job, put off estimating the final load until after you have finished pouring from the fully loaded trucks. You won't have a lot of time in which to make that final calculation in order to send it back with the last driver or call it in to the dispatcher, but you will have cut down your margin of error by dealing with only the few yards that remain rather than with the whole job.

On the day of the pour, leave yourself enough time for careful figuring and double-checking. Last-minute concrete panic has a long tradition, but punching the keys on a pocket calculator while you are trying to get the building inspector to sign off the formwork will make for miscalculations every time. Have a helper figure independently how much concrete you'll need, and then compare notes. This should pick up careless errors in math, and the easily made mistake of forgetting to figure in some portion of the pour.

Calculating volume—The only accurate way to calculate most residential concrete jobs is to get down in the trenches and take measurements. Write down the width, depth and length of each trench or form. Having to remove a large root from a footing trench and undersupervising an overeager backhoe operator are just two reasons why the actual measurements may differ from the numbers on your blueprints enough to make a real difference.

Particularly with slabs and trenches, the depth figure can be a compromise based on measurements taken at many points and tempered by the kind of intuition that comes with experience. The same can be true of trench width, depending on the soil.

Once you've noted the measurements for each wall, footing or slab, group them into separate categories, one for each configuration or cross section. If the depth of a footing trench changes, figure this portion of the footing as a separate category, even if the width remains the same. When a configuration has more than four sides, such as a T footing, break it down into separate rectangles, trapezoids or triangles. Keep all this information on a clipboard. I number the categories, draw a small section of each one and fill in the dimensions so I don't get confused. List the different lengths for each cross section below the appropriate drawing, and note its location on a plan drawing of the pour. This really helps later in the pour when you have to calculate the last load in a hurry. You can now add the lengths of all the footings and walls in each category to get the total lineal feet for each configuration.

Concrete is ordered by the cubic yard. Unfortunately, the width and depth of footings, walls and slabs are often in inches, and length is in feet and inches. Calculate each cross section—width x depth—in inches if the increments are small, or in feet and decimal feet if the numbers get too big. Just don't mix the two. To convert your square-inch answers to square feet, divide by 144. List the total area of each cross section on your clipboard. Use a pocket calculator to grind out these numbers.

To get the total concrete needed in each category, multiply the cross-section figure, now in square feet, by the total of all the lengths in the category, which is already in feet. The product will be in cubic feet. Adding all of the totals for the categories together, and dividing by 27—the number of cubic feet in a cubic yard—will yield a total in cubic yards.

You can order concrete in fractional yards, but remember to round off high, not low, to get to the nearest large fraction. I usually order a few extra cubic feet to protect against being short, in addition to the standard 5% allowance for spillage. It's a good idea to have pier holes for decks or retaining-wall footings dug to use any excess concrete.

Placing your order—If there is more than one batch plant near the job site, ask around to see which one other builders like. Since price per yard is usually similar, their impressions will be based on phone contact with the dispatcher, and on pouring with the drivers. These opinions are useful, because the cooperation and expertise of the people in these two positions will determine how easy and successful your pour will be.

Give the dispatcher the day and time you want to see the first concrete truck, and how many yards you'll need. Then describe the mix you want, in enough detail so that the proper concrete will get sent to your job site. There are two established methods— performance and prescription.

When you order with a performance specification, you give the dispatcher a compressive strength in psi, and it's up to the batch plant to supply concrete that will test to that minimum figure in 28 days. Prescription ordering, on the other hand, puts the responsibility on you. It also lets you determine some of the variables in the mix design. You may want to duplicate a mix that worked well for you in the past, or to satisfy a restriction unique to this pour, such as a maximum aggregate size that the pumper can handle.

A minimum prescription tells the dispatcher how many pounds or bags of cement to use with every cubic yard of concrete. When you specify only the cement content, the batch plant will determine all the other variables.

If you are knowledgeable, or if there are engineering specifications on your plans you need to satisfy, carry prescription ordering a step further by specifying slump, maximum size of the coarse aggregate, or the percent of air-entrainment. If not, you should mention any special characteristics of the site or of the pour that will affect batching or delivery. If weather is a problem, ask about admixtures. If you are pouring grade beams on a steep slope, talk about slump. If you are going to use a pump, tell the dispatcher which company, confirm the time of the pour, and have the aggregate and mix adjusted.

Whatever way you order, make sure you get a batch ticket from the driver for each load you receive. This is more than just an invoice that lists the number of yards of concrete. It should also tell you the cement and water content of the mix, the size and amount of aggregates, the amount of air-entrainment, the percentage or weight of admixtures, and the slump.

Maybe most important, give the dispatcher clear directions to your job site and a telephone number where the driver or the batch-plant dispatcher can reach you or someone who can get in touch with you. It's surprisingly easy to lose a truck that weighs 27 tons, but it's more than difficult—and expensive—to deal with its load after it's sat in there a few hours.
—*Paul Spring*

Curing Concrete

Keeping it warm and wet for a week
will make concrete stronger and more watertight

by Bruce A. Suprenant

Some builders think that after the concrete is placed—whether in a wall or in a slab—the job is over. They don't understand that poorly cured concrete is weak and porous, and more likely to be damaged by cracks, wear, rebar corrosion, chemicals and freeze/thaw cycles. Improperly cured concrete may reach only half of its specified strength (see graph below). Smart builders, on the other hand, postpone Miller Time for a while and take steps to keep concrete moist—curing until it reaches its full strength. They also look for better, easier and cheaper methods of concrete curing.

A small portion of the water added to the concrete mix combines chemically with the cement in a process called hydration. Unfortunately, most of the water added to the concrete mix evaporates and leaves voids. These voids represent the difference between concrete that is properly cured and concrete that is poorly cured or not cured at all. Good curing enables the hydrated cement to occupy most of these voids, and thereby increase the strength and durability of the concrete.

Temperature and time—Temperature affects the speed of hydration. The warmer the concrete, the faster cement hydrates and gains strength. At air temperatures between 50° F and 90° F, there's no need to control concrete temperatures. Fortunately, much concrete is placed within this temperature range, so a builder does not usually have to spend money on temperature control. Outside this range, though, a builder may have to take steps to heat or cool the concrete.

Concrete temperatures below 50° F make it hard to achieve early strength. Hydration is slow, so early strength is low. Below 40° F early-strength development is greatly retarded, and at 32° F little early strength develops.

Cement hydrates faster at temperatures above 90° F, and concrete gains strength rapidly. Its ultimate strength, however, isn't as high as that of concrete cured at a lower temperature (see graph, facing page). Also, later thermal cracking is more likely to occur in concrete placed on a hot day followed by a cool night. Thermal cracking looks like regularly spaced cracks running all the way through the concrete and is caused by changes in temperature.

According to the American Concrete Institute's "Specifications for Structural Concrete

Curing compounds seal the surface of the concrete, preventing water from evaporating. In the photo above, a worker uses a garden-type sprayer to apply curing compound to a stamped concrete patio. *Photo courtesy of Concrete Construction magazine.*

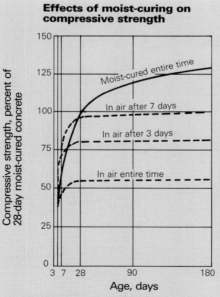

Effects of moist-curing on compressive strength

Compressive strength, percent of 28-day moist-cured concrete

150
125 — Moist-cured entire time
100 — In air after 7 days
— In air after 3 days
75
50 — In air entire time
25
0
3 7 28   90   180

Age, days

In all cases the rise in compressive strength is dramatic over the first seven days. Beyond that, the advantages of moist-curing are obvious, resulting in concrete where compressive strength is anywhere from 25% to 75% more than concrete that is simply air-dried.
(Copyright 1983 National Association of Home Builders)

for Buildings" (see sidebar, p. 16) curing must continue for at least seven days for all concrete, except high-early-strength concrete (concrete made with Type III cement). At least three days of curing are required for high-early-strength concrete. You may be able to shorten the curing period, but only if you run strength tests on cylinders that have been kept adjacent to the pour and cured by the same methods. You can stop curing when the average compressive strength of these field-cured cylinders reaches 70% of the specified strength. The specified strength at 28 days is usually 3,000 lb. per square inch (psi), but check with your concrete supplier to verify the strength for your mix.

In addition to making efforts to retain moisture in the concrete, maintain the temperature of field concrete at a minimum of 50° F for as long as the laboratory-cured cylinders take to reach 85% of the specified strength. This temperature/time requirement should range from 7 to 12 days for all concrete, except for high-early-strength concrete, which ranges from 3 to 7 days. Once again, check with the supplier for a time estimate of curing for your concrete.

Cylinder testing—Don't let the idea of cylinder testing scare you. Most cities have two or three testing firms that will sell you plastic cylinder molds and will test the cylinders after you make them. Cylinders 6 in. in diameter and 12 in. long cost about $1 each (buy 6 to 8 for each job to be tested), and testing usually runs $10 to $15 per cylinder.

To make the cylinders, place the concrete with a scoop or trowel in three equal layers within each cylinder. After placing each layer, use a ⅝-in. dia. rod to mix the concrete by moving the rod up and down 25 times (called rodding). After the final layer has been rodded, screed off the top to form a level surface and lightly tap the sides of the cylinder with your hand to expel small air bubbles. Cure these cylinders using the same methods you use for the structure.

For most concrete, wait about five days and then take two cylinders into the testing lab. The test lab will break the cylinders and provide you with the strength results on the same day that you bring them in. If the strength results are not equal to 70% of the specified concrete strength, curing must continue, and another test will be required. Check with your concrete supplier for the approximate time when 70% strength will be achieved. This information can minimize the number of cylinders purchased and made, trips to the test lab, and overall cost. Testing with cylinders may cost up to $100, but may save two or three days in curing. This can be important on big jobs, but if time is not important, simply cure the concrete for seven days.

Curing methods and materials—The methods for maintaining the moisture content needed for curing fall into two categories. Applying water continuously calls for ponding water on slabs; sprinkling or fog-spraying; using soaker hoses and applying water-saturated cover materials such as burlap, straw, earth or sand. The second method calls for preventing excessive moisture loss, either with curing compounds or by covering the concrete with polyethylene sheets (photo right) or with reinforced waterproof paper. Most of these items are readily available at the local lumberyard and building-supply store. Additionally, most concrete suppliers either sell or know where to buy them.

Applying a curing compound to the concrete forms a membrane that seals the water into the concrete. This membrane is not completely waterproof, however, and a small amount of moisture is lost through evaporation. In this respect curing compounds aren't as effective as waterproof paper or polyethylene sheets, which retain all the moisture if they remain intact and in place during the entire curing period. But curing paper or plastic may be damaged on a site or dislodged by wind or workers.

Curing compounds are easy to apply on both horizontal and vertical surfaces, and for many concrete jobs, they are the most economical curing method. But if the concrete is to be coated, painted or covered with a flooring material, make sure that the curing compound won't interfere with the bonding of the final finish. Most curing compounds do interfere with bonding.

Generally, curing compounds are best applied right after finish-trowelling. They cost between $4 and $6 per gallon, and the typical coverage is 200 sq. ft. per gal. Spraying is the fastest method and typically, a farm or garden spray can is used (photo facing page). To ensure proper coverage, it's best to use two applications at right angles to each other. Brushes and rollers will work for small areas or small budgets. Pigmented compounds, colored white or grey, make it easier to see if the material has been applied uniformly. This helps eliminate missed spots and pinholes that can leak excessive moisture through the curing compound. The color fades away after a few days.

If you're dealing with colored concrete, use the curing methods recommended by the manufacturer of the dry-shake color or coloring admixture. Color-matched curing waxes are widely used for flatwork. Many

While curing compounds permit some evaporation, covering a slab with polyethylene sheeting will retain all of the moisture in the concrete. The disadvantage is that the poly can easily be damaged by wind or workers. *Photo courtesy of Portland Cement Association.*

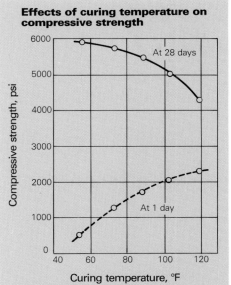

Effects of curing temperature on compressive strength

Increasing the curing temperature increases the early strength of concrete, but decreases final strength. For the material shown in the graph, 1-day strength increased from 800 psi to 2,000 psi when curing temperatures increased from 60° F to 100° F. But 28-day strength decreased from 5,900 psi to 5,200 psi. (Courtesy of the Portland Cement Association)

curing methods may stain the concrete or cause efflorescence (white surface deposits). Plastic sheets and waterproof paper can cause a blotchy appearance because of uneven moisture distribution. Even curing compounds used for normal concrete are likely to be unsightly.

Curing in cold weather—Avoid curing with water during freezing weather. Water running out of heated enclosures freezes and may cause an icing hazard around the job site. Also, water-cured concrete is likely to be saturated when curing has ceased, which makes it vulnerable to damage from freezing.

In cold weather use curing compounds, polyethylene sheets, or waterproof paper to retain moisture. Then as hydration proceeds, internal water-filled voids will be partially emptied. When compressive strength reaches 500 psi (for most mixes this occurs at about 24 hours) water in the voids will be reduced enough to prevent damage from freezing.

For both temperature protection and moisture retention, you can cover the concrete with thermal curing blankets, available from the following companies: Acme Canvas Co., Inc. (171 Meford St., Malden, Mass. 02148); Ametek Microfoam Div. (Rtes. 1 and 202, Brandywine 4

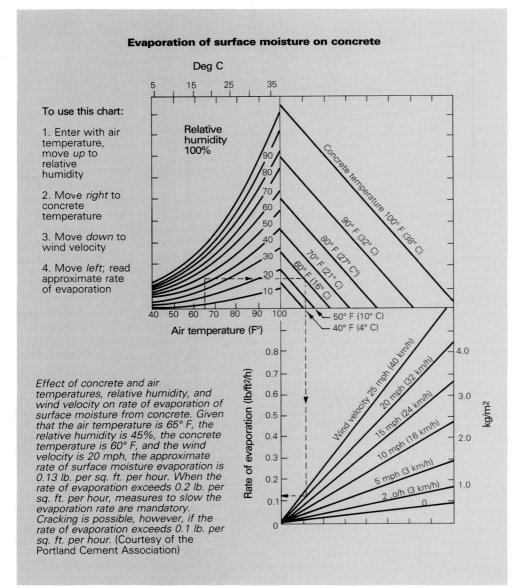

Evaporation of surface moisture on concrete

Deg C

To use this chart:

1. Enter with air temperature, move *up* to relative humidity

2. Move *right* to concrete temperature

3. Move *down* to wind velocity

4. Move *left*; read approximate rate of evaporation

Relative humidity 100%

Concrete temperature 100° F (38° C)
90° F (32° C)
80° F (27° C)
70° F (21° C)
60° F (16° C)

Air temperature (F°)

50° F (10° C)
40° F (4° C)

Wind velocity 25 mph (40 km/h)
20 mph (32 km/h)
15 mph (24 km/h)
10 mph (16 km/h)
5 mph (3 km/h)
2 o/h (3 km/h)
0

Rate of evaporation (lb/ft²/h)

kg/m²

Effect of concrete and air temperatures, relative humidity, and wind velocity on rate of evaporation of surface moisture from concrete. Given that the air temperature is 65° F, the relative humidity is 45%, the concrete temperature is 60° F, and the wind velocity is 20 mph, the approximate rate of surface moisture evaporation is 0.13 lb. per sq. ft. per hour. When the rate of evaporation exceeds 0.2 lb. per sq. ft. per hour, measures to slow the evaporation rate are mandatory. Cracking is possible, however, if the rate of evaporation exceeds 0.1 lb. per sq. ft. per hour. (Courtesy of the Portland Cement Association)

They are the result of rapid moisture evaporation caused by high temperatures, low relative humidity and wind greater than 15 mph. The graph at left illustrates a method for evaluating the possibility of plastic shrinkage cracking. If the surface evaporation rate exceeds 0.2 lb. per sq. ft. per hour, plastic shrinkage cracking is likely to occur.

To minimize the possibility of plastic shrinkage cracking, take the following precautions: moisten subgrade and forms; erect temporary windbreaks and sunshades; place the concrete during the cool time of the day; reduce time between placing, finishing and start of curing. Immediate, continuous curing during hot weather also helps to prevent crazing (eggshell or chicken-wire patterns), cracks which are usually caused by early drying or intermittent wetting and drying.

Curing inexpensively—Under most conditions, leaving wall forms in place is an acceptable curing method. Keeping the forms wet helps to cool them, further reducing moisture loss. Builders should weigh the cost of leaving forms in place for up to seven days against stripping forms as soon as possible and curing by other methods.

Absorptive wooden forms left in place are not a satisfactory means for curing structures during hot, dry weather. Loosen the forms as soon as possible so curing water can be run down inside them. According to the American Concrete Institute, natural curing from rain, mist, high humidity, low temperature or moist backfill is sometimes sufficient. Its effect must be at least equivalent to keeping the concrete above 50° F and moist for the first 14 days with Type II cement, 7 days with Type I cement, or 3 days with Type III cement.

Curing properly—Unfortunately, no field test lets you know if the concrete was cured properly. Strength tests of field-cured cylinders can indicate curing effects, but these cylinders are also affected by vibration, their size and other factors. Also, poor curing has its most pronounced effect on the thin layer of surface concrete. The lower quality of this layer might affect the strength only marginally, but may reduce durability considerably.

Color differences give a clue to curing effectiveness. Well-cured concrete is typically darker in color than poorly cured concrete. But color differences are also caused by changes in cements and admixtures, so color comparison is a crude method at best.

The best way to guarantee proper curing is to choose the right method, initiate it promptly, follow recommended applications and continue curing for the required time period. With a little tender loving curing, your concrete will reward you with many years of good performance. □

Bruce A. Suprenant is an engineer who teaches at the University of Colorado at Boulder. He is a contributing editor of Concrete Construction *magazine.*

Bldg., Chadds Ford, Pa. 19317); Raven Industries, Inc. (Flexible Films Dept., P. O. Box 1007, Sioux Falls, S. D. 57117); Reef Industries (P. O. Box 750250, Houston, Tex. 77275-0250).

Curing in hot weather—During hot-weather concreting operations, water evaporates from the concrete, so water curing is preferred. Effective methods include placing wet burlap on the concrete and covering it with polyethylene or spraying it continuously with water. For large areas of flatwork, curing compounds are more practical. White-pigmented curing compounds help reduce concrete temperatures because they reflect sunlight and reduce heat absorption.

Rapid drying of flatwork in hot weather causes surface cracking. Workers have to apply water, curing compounds or coverings quickly after finishing to prevent plastic shrinkage cracking. If final curing is delayed, you can try fog-spraying as an intermediate curing method to prevent or reduce the chances of cracking.

Plastic shrinkage cracks are relatively short (2 ft. or 3 ft.), shallow (½ in. to 1 in.) parallel cracks that may occur before final finishing.

Publications on curing concrete

American Concrete Institute (P. O. Box 19150, Redford Station, Detroit, Mich. 48219)
"Standard Practice for Curing Concrete" (ACI-308-81, rev. 1986), $6.25
"Hot Weather Concreting" (ACI-305R, revised 1986), $10.50
"Cold Weather Concreting" (ACI-306R, revised 1986), $13.95

National Association of Home Builders (15th and M Streets, N. W., Washington, D. C. 20005-4099)
"Residential Concrete," $12.00

Portland Cement Association (5420 Old Orchard Rd., Skokie, Il. 60077-4321)
"Design and Control of Concrete Mixtures" (EB001T), $19.50 plus $2.00 postage
"Cement Mason's Guide" (PA122) $4.75 plus $2.00 postage

Site Layout

On a flat lot, footings can be oriented with precision using batter boards, string and a water level

by Tom Law

Driving that first stake is always exciting. It doesn't matter whether you've been designing and dreaming about the house for years, or you're beginning the first day of actual work on what you hope will be a profitable contract. Laying out the site in preparation for foundation work is your first chance to visualize the house full scale in its setting. The accuracy of your layout and the foundation it defines will also determine how much you will have to struggle to make your house tight and square.

Unless you are building on a sloped site, all you will need to do the job right is a 100-ft. tape measure, a water level, a ball of nylon string, enough lumber for batter boards, a helper and the application of some practical geometry.

Let's assume that the house is a simple rectangle, and that one of the long walls faces south. If precise solar orientation is important, use an accurate compass or one of the many siting devices available commercially. If such precision is not necessary, just stand facing the midday sun. Your outstretched left arm will point to the east and your right arm to the west. Unroll a ball of string along this axis for a distance a few feet greater than one of the long

walls of your building. I use braided nylon string because it will take an awesome amount of tension before it breaks, and because braided string doesn't unravel and can be used indefinitely. It comes in a highly visible yellow as well as in white.

Preliminary layout—Select one end of the layout as a starting corner and drive a small stake into the ground. You can use almost anything for a stake—a timber spike or a tent peg—as long as it holds the string off the ground. Now measure back down the string the length of the wall, drive another stake, stretch the string between the two stakes and tie it off. This lets you adjust the placement of the house on the site, and gives you an idea of where to locate the batter boards. Taut strings and accurate squaring are not necessary at this point, as long as the outside dimensions of the house are accurate.

To lay out the rectangle, pull another string from the corner you just established, at a right angle to the long wall. This is where your helper is needed. One person pulls the string while the other guesses at 90°. Measure along

this second string the width of the house and drive the third corner stake. The fourth corner can be found by measuring. Now step back and study the house placement on the site. Stop and think on this one a while—it's a permanent decision. If you are satisfied, you can begin preliminary squaring.

With a 100-ft. tape, measure the diagonals. If they measure the same, then you have created a rectangle. If not, you have a rhomboid, and you will have to adjust the two corner stakes opposite the long, south wall until the diagonals are about equal. Getting to within one or two inches at this point is close enough. I use a steel tape because cloth tapes stretch. A leather thong tied to the metal loop on the zero end of the tape will help you to pull hard and hold a dimension at the same time.

Crouch with your forearms braced against your thighs, and use your body weight to pull against the person on the other end of the tape. On the zero end, hold the leather strap, not the tape, and when you are squaring strings that are suspended from batter boards later on in the layout, keep the tape from lying on the string and deflecting it. Another method of get-

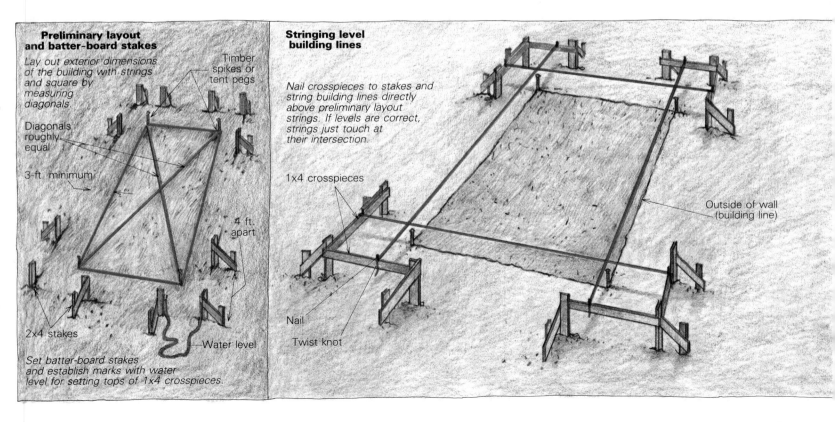

Preliminary layout and batter-board stakes

Lay out exterior dimensions of the building with strings and square by measuring diagonals.

Timber spikes or tent pegs

Diagonals roughly equal

3-ft. minimum

4 ft. apart

2x4 stakes

Water level

Set batter-board stakes and establish marks with water level for setting tops of 1x4 crosspieces.

Stringing level building lines

Nail crosspieces to stakes and string building lines directly above preliminary layout strings. If levels are correct, strings just touch at their intersection.

1x4 crosspieces

Nail

Twist knot

Outside of wall (building line)

ting an accurate reading on a tightly stretched tape is to hold the 1-ft. mark rather than the zero. This allows you to grasp the tape with both hands when holding it over a string intersection. With this method, remember to tell the person on the other end of the tape that you are "burning a foot," or "cutting a foot," so the measurement can be adjusted accordingly.

Once this preliminary layout is approximately square and located where you want it on the site, you can set up batter boards. Batter boards are fixed in the ground out beyond the excavation lines. They are temporary wooden corners used to tie the string that accurately defines the perimeter of the building at the outside-of-wall line, or *building line*, and the outside-of-footing line, or *excavation line*. If excavation is required within the perimeter of the structure, for an interior footing or a line of piers, you may want to establish batter boards to hold strings for these lines as well.

I use 2x4 stakes, 3 ft. to 4 ft. long, with 1x4 crosspieces. Usually this is lumber that has been used at least once before. Sharpen the stakes with a circular saw so they will drive easily with a sledgehammer. You'll need three stakes per corner, set about 4 ft. from each other, for a total of 12 for the rectangular house we're using as an example. Drive these stakes about 3 ft. outside the preliminary strings and parallel to them. This placement gives you enough room so that the excavation won't undermine the batter boards, and they can be used until the walls of the house are actually framed. It's a temptation to be exacting in placing your stakes, but you needn't take the time to be too fussy. A good foundation requires precision, but this comes from the strings the batter boards will eventually carry, not from the batter boards themselves. Be sure that the 2x4 stakes are rigid enough to withstand an occa-

sional bumping. If they aren't, nail 1x4 braces near the top of the 2x4 stakes, drive another stake where the brace touches the ground, and nail them together.

Stringing level building lines—I like to nail all the crosspieces at the same level whenever possible. When the strings are in place, this gives me a vertical reference anywhere on the perimeter, which is a real advantage in determining the depth of footings, or the height of foundation walls and concrete block. Since a lot of my foundations are block, I like to set my crosspieces (where the strings will eventually be tied) so that their top edges will be at the same height as the top course of block. To figure this, you must start at the bottom of the footings. In the colder parts of the country, the bottom of footings must always be at or below the frost line to prevent heaving during the winter. If the frost line is 32 in., for instance, and the depth of the footing itself is 8 in., then it will take 24 in., or three courses of 8-in. block to reach grade. Add another three courses, as a convenient and attractive foundation height, and you have a total of 56 in. from the bottom of the footing trench. Keeping this number in mind, measure 24 in. up from the ground on any of the batter-board stakes—the finished height of the block foundation—and make a level line.

I don't own a builder's level or transit, and I've never needed one in my 20 years in the trades; I use a water level (you can make your own). Whatever you use, mark level lines on all the batter boards at 24 in. Accuracy is very important here. Then nail the 1x4 crosspieces with their top edges even with the level marks.

When the crosspieces are up, pull a new string for the south side of the building over the crosspieces of batter boards on each end. Align

it directly above the preliminary layout string by sighting it from above or using a plumb bob. Tack a nail in the top of the cross member and tie one end securely. At the other end pull the string as tight as you can. This establishes a line of elevation, so you don't want it to sag. Use a twist knot to tie it off to another nail. The twist knot (drawing, below right) will keep a nylon string taut, while still allowing it to be released instantly for resetting. This knot doesn't work well with cotton line. Continue stringing until all the lines are up. If everything is level, the strings will just touch as they intersect a few feet in front of the batter boards.

Squaring the corners—The next step is to square the stringed corners, this time using the 3-4-5 check. These numbers refer to the sides and hypotenuse of a right triangle. Since 6-8-10 and 12-16-20 triangles are proportional to a 3-4-5, use the largest one you can for optimum accuracy. The intersection of the strings of each corner defines the 90° angle of the 3-4-5 triangle. You'll need a helper to measure and adjust the strings until the hypotenuse is exactly proportional. With one person holding zero (or 12 in., if you are "burning a foot") on the tape, the other person can mark the legs of the triangle on the string with a pencil, and then knot a short length of string loosely around the mark. Double-check the measurement, and then tighten this knot. Measure from knot to knot to get the hypotenuse, and adjust the strings on the batter boards if necessary. These adjustments will require driving new nails into the top edge of the crosspieces. Pull the previous nail as you correct the position of the string. If you don't, it can get very confusing when the strings come down temporarily for digging the footing trenches. When you take the string down and put it back up, check

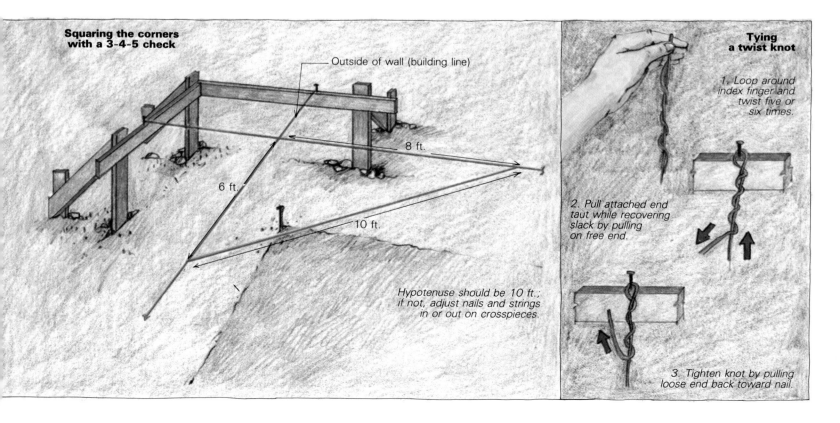

Squaring the corners with a 3-4-5 check

Outside of wall (building line)

8 ft.

6 ft.

10 ft.

Hypotenuse should be 10 ft.; if not, adjust nails and strings in or out on crosspieces.

Tying a twist knot

1. Loop around index finger and twist five or six times.

2. Pull attached end taut while recovering slack by pulling on free end.

3. Tighten knot by pulling loose end back toward nail.

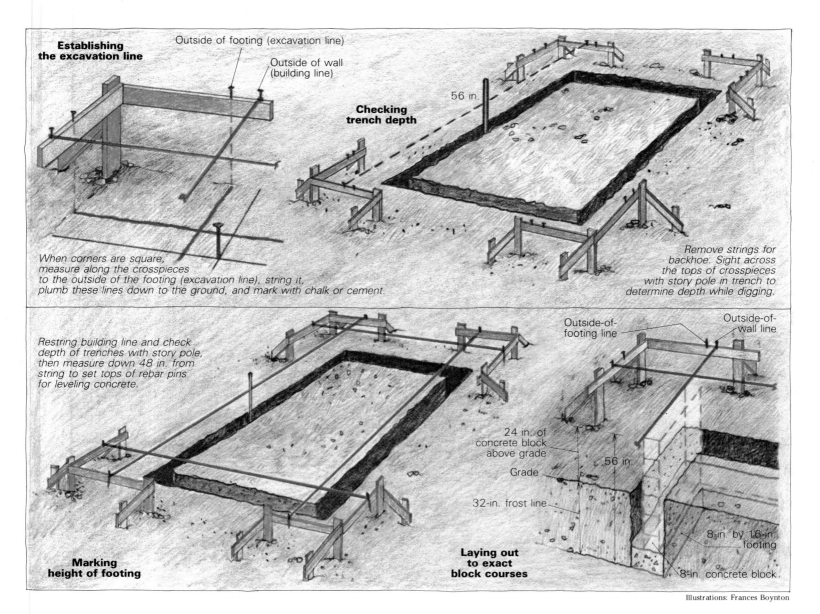

Establishing the excavation line

Outside of footing (excavation line)

Outside of wall (building line)

When corners are square, measure along the crosspieces to the outside of the footing (excavation line), string it, plumb these lines down to the ground, and mark with chalk or cement.

Checking trench depth

56 in.

Remove strings for backhoe. Sight across the tops of crosspieces with story pole in trench to determine depth while digging.

Restring building line and check depth of trenches with story pole, then measure down 48 in. from string to set tops of rebar pins for leveling concrete.

Marking height of footing

Laying out to exact block courses

Outside-of-footing line

Outside-of-wall line

24 in. of concrete block above grade

Grade

56 in

32-in. frost line

8-in. by 16-in. footing

8-in. concrete block

Illustrations: Frances Boynton

the length to the knots, because nylon string stretches. To finish squaring up, use the 3-4-5 check on another corner, then check the diagonals again.

With the strings squared up to represent the eventual building lines (the outside-of-wall lines), and an elevation established, draw a plumb line down from the strings on the face of each cross-member, and write the wall thickness and the amount that the footing will project beyond the outside of the wall on the batter board. To reduce confusion, I drive nails and hang strings only on the outside-of-wall line and outside-of-footing line.

Excavation lines—Usually a footing is twice as wide as the foundation wall it supports, and as deep as the wall is thick. Footings are contained either by building a wooden form, or by digging a trench and using the undisturbed earth as formwork. I usually use the trench method. To show the backhoe operator where to dig, I plumb down to the ground from the outside-of-footing line and stretch a string at grade. I mark over this string with lime or cement dust as if I were marking out an athletic field. You can also use scouring powder with a shaker top. The backhoe operator should hold the outside tooth of his bucket to the line, and dig to the inside.

Checking trench and footing depth—The batter boards give a quick vertical reference for determining how deep to trench. In our example, the bottom of the footing is 56 in. down from the top edge of the crosspiece. Instead of strings, which during excavation should be wound around a stick, use a story pole with a 56-in. mark on it. Stand the stick in the trench and sight from the top of one batter board to another. The mark on the story pole should line up with them.

When the machine work is finished, string the lines on the outside-of-wall line (the building line). Pull them very tight. Shape up the sides and corners of the trench with a shovel, maintaining the 56-in. depth you can now check by measuring from the string.

Next, set the depth of pour for the concrete. I use ⅜-in. or ½-in. steel reinforcing rods about 6 ft. to 8 ft. apart to indicate depth during the pour. Cut them about twice the depth of the footing so that you can drive them into the ground. Measuring down from the string to the top of the rebar, carefully tap them with a sledge until you read 48 in. on the tape. This

will give you an 8-in. footing at the 32-in. frost line, and six courses of block on top will bring you up to the string (drawing, above right). Then pour the concrete level with the top of the rebar. I use a garden rake to push the concrete around and for initial screeding. I hold the rake in a vertical position to smooth the top of the concrete and jitterbug the coarse aggregate down into the mix. You also might want to use a 2x4 screed short enough to fit between the rebar depth indicators, but it's not necessary to trowel the surface smooth. If you are pouring a foundation wall on top of the footing instead of laying block, the same techniques can be used, but remember to form a keyway in the footing to receive the next pour, and check with local codes to see if vertical rebar is required to tie the footing to the foundation wall.

The next day the concrete will be hardened sufficiently to begin working on the foundation walls. I usually drop a plumb bob down from the outside of wall lines and snap chalklines on the green footing. If the foundation is to be concrete block, then marking the corners will be enough since the mason will be pulling his own lines from corner to corner on each course. □

Tom Law is a builder in Davidsonville, Md.

Stemwall Foundations

Use plywood, framing lumber and snap ties to form the foundation, then build the house out of the forms

by Dan Rockhill

I poured my first concrete foundation on a site tangled with mature, unpruned trees. Furthermore, we were laboring under the watchful eye of an uneasy client. Initially I had no intentions of building the foundation, but our subcontractor threw us a curveball when he suddenly announced that there would be a two to four-week delay before he could pour the footings. As I pondered the delay and our anxious client, I decided to do it myself. My crew and I will never forget that decision.

The site had bedrock fingers that came almost to the surface, making it difficult to drive stakes into the ground. We were pouring a grade beam that would sit directly on the rock, and because it was next to impossible to penetrate it with a stake we stabilized the forms with an elaborate network of crisscrossing braces,

which rendered the interior side of the foundation virtually inaccessible. But this condition didn't concern us, as we were going to let the concrete "seek its own level." All we had to do, according to my friend Richard, was to add plenty of water to the concrete.

It did indeed seek its own level. Soupy concrete oozed from the tiniest holes in the forms, and flooded into the future crawl space through cavernous gaps between the bottom of our forms and the irregular bedrock. To the utter disbelief of the first concrete driver to arrive, we watched the first two or three yards pour into, and then dribble out of, our forms.

Now one thing I did know about contracting, from all the books I had read, was that you were supposed to get the dirt piles as far away from the excavation as possible. This we had done

with diligence during our site prep, not realizing that it would be the only material around to patch the holes between grade and our forms. We started running what seemed like the high hurdles, carrying shovelfuls of dirt from one side of the site to the other. Even the concrete driver pitched in to help, and I know now from experience how unusual that is.

Six hours later, when the last concrete truck left, we collapsed on the grass only to be accosted by our client. She was in a state of delirium because the concrete truck had broken a few twigs on her maple trees as it jockeyed around the site.

Since that dreadful day I have worked on refining a simple system of concrete forms that I can assemble with the help of the most inexperienced crew. The key parts of this system are plywood, 2x4s and an ingenious device called a snap tie (drawing, right).

A snap tie is a slender metal rod that connects the opposing walls of a form, holding them apart and locking them together at the same time. Once the concrete is in place, you break off the protruding metal stem and pry out the plastic cone. The resulting crater is easily patched if you're concerned about appearance. Snap ties cost about $.30 apiece, but they ensure accurate work and they are so much easier to use than wood spacers that I wouldn't build a foundation without them. We used about $60 worth of snap ties on the relatively small foundation shown in the photos. We have our own wedges that are used in conjunction with the ties. They cost a nickel apiece to rent, or they can be bought for about $.70 apiece.

I assemble form panels using full sheets of plywood, which are perfect for building 4-ft. high stemwalls atop a separately poured footing. Commonly known as a T-wall system or crawlspace foundation, this kind of foundation is popular in areas with a medium depth of frost penetration (28 in. to 42 in). We use them here in Kansas, where we have to excavate too deep for a slab foundation but not deep enough for a full basement. The resulting 4-ft. high crawl space has the added benefits of being large enough to house mechanical systems and of serving as a handy tornado shelter. Having all the mechanical services in the crawl space frees up valuable square footage above that might otherwise be lost to hot-water tanks or furnaces.

Footings on a vacant site—Most building projects begin with nothing more than a stake driven into the ground to locate a corner of the structure for reference. I always feel a little reverential at this moment, because this act symbolically marries me to the project and quite deliberately initiates construction.

There are as many different kinds of sites as there are buildings, so it is difficult to give any hard-and-fast rules for excavation. Typically, you begin by carving away enough soil for a generous crawl space. Then you should lay out and mark the footings (for details on foundation layout, see Tom Law's article, "Site Layout," on pp. 17-19). Mark the outlines of the foundation trenches with lime so that your backhoe driver can easily see where to cut. Don't use bonemeal to mark the lines. I did once, and the client's dog licked it up as soon as I left the site.

The minimum footing depth is determined by the maximum depth of frost penetration. If you don't know what it is, check with the local building-inspection office. Note that the minimum footing depth can be affected by the quality of the soil. Some soils hold up buildings better than others. Consult with a soils engineer if you have any doubts about the dirt.

A good backhoe operator can save you a lot of headaches by cutting straight trenches to the desired depth. Ask around to find out who the good ones are, and don't base your selection strictly on hourly rates.

Before the backhoe arrives, decide where you want to put the dirt that comes out of the trenches. You need to save some for backfill,

In stable soils, a footing can be poured directly into a trench cut into the earth. But in crumbly soil like this, Rockhill uses framing lumber to make forms for the footings (photo above). Their tops are held together with snap ties. Their sides are bolstered with stakes and tamped earth.

and if you have some good topsoil, set it aside for finish grading. Before digging, be sure that you know where any underground utilities are located. If you don't, I can assure you the backhoe will find them and snap them in a split second. This creates unnecessary delay and danger that can easily be avoided. All utility companies are pleased to come out and mark the positions of their underground lines.

In some parts of the country the soil is stable enough to allow you to "trench pour" a footing. This means that all you do after digging a clean trench is to set the steel, then some grade stakes to control the footing height. Then you can fill the trench with concrete. In soils around here, it's hard to get a perfect trench so I use wood to form the footings (photo above). The side walls of our trenches have to slope back enough to prevent a cave-in, and the trenches have to be wide enough to muscle forms around in.

When the trenches are complete, be certain

that you leave the soil undisturbed beneath the footing. If any soil has been removed and replaced, it must be tamped solid. Around here we can rent gasoline-powered compactors, or "jumping jacks," for about $80 a day.

In most situations, a T-wall footing should be twice as wide as the wall is thick and as deep as the wall is thick. For a standard 8-in. thick stemwall, that means a footing that measures 8 in. by 16 in. I form all the footings out of the dimension lumber that will later be used as joists or headers. When you begin to use the lumber later, be sure to have plenty of sawblades on hand. The concrete dust dulls them quickly. The forms for the footing don't have to be immaculate, as long as they are sturdy and the size of the footing is not compromised. Boards can be lapped or run long at the outside corners in order to minimize cutting.

If I'm using 2x10s for the footing forms, I find it easiest to pour the concrete to the top of the form. On small foundations, the expense of the extra 1¼ in. of concrete is offset by the ease of this operation. If I'm using 2x12s, I snap a chalkline at a height of 8 in. inside the forms and set a few nails along the line as reference points because the concrete will likely obliterate all traces of the chalk.

I have found that the entire job goes a lot faster if I take the time to make my footings dead level. This means carving away obstacles at the bottom of the trench, and making plenty of level checks with the transit. But I think the extra work is worth it because it sets a good precedent for the rest of the job.

To keep the forms at the right level, I drive stakes a couple of feet into the ground and nail through them into the side walls of the form with 16d duplex (double-head) nails. Duplex nails are a lot easier to remove when it comes time to strip the forms. I try to set my stakes about every 4 ft. I also tie the opposing form boards together with a 2x4 collar every 4 ft. or so. The collars impede the pour somewhat, but can be navigated around.

If the site slopes, make sure that the footing stays below the frost line. To do this you will have to step the footing down the slope (stepped foundations look like staircases stretched across the lot). So that you don't weaken the footing, make sure the concrete is at least 12 in. thick where it steps down.

Patch gaps between the stepped forms and the grade with tamped earth before you place the concrete. You can make the "treads" of the step footing as long as you like but not less than 24 in. A 4-ft. tread is good because it maximizes the use of your plywood.

Rebar—Steel gives a backbone to concrete and is a necessary part of your footing. Most 16-in. wide footings use two continuous ½-in. reinforcing bars spaced 8 in. apart down the center of the footing, directly below the inside and outside lines of the future stemwall. The steel will resist tension forces better than the concrete will alone. Rebar comes in 20-ft. lengths, and where sections of it meet they should overlap by at least 12 in. (24 rebar diameters). The overlaps should occur in straight runs of foundation—never at the corners—and should be staggered in parallel runs. Thoroughly secure adjoining pieces with form wire to keep them from working free as the concrete is placed.

To cut rebar you can use either a reciprocating saw fitted with a metal-cutting blade or a hacksaw to cut partway through the steel. Then bend it over your knee to snap the pieces apart. It's easier, though, to use a cutting torch or a cutoff blade in a circular saw. The best cutting method is a rebar cutter/bender. This tool costs about $250, but if you do a lot of rebar work you can easily justify the cost.

Many builders support the rebar in the footing with bricks or broken chunks of block placed about every 10 ft. This holds the steel the required minimum 3 in. above the bottom of the footing, but does not hold it securely in place. I tie rebar to the forms with form wire. To do this, drill two ¼-in. holes, 3 in. and 4 in. up from the bottom of each form board after it is placed,

and thread through some form wire. Weave the wire around the steel (photo facing page, top left) and tie it off outside the form. Then use a nail to twist the wire around the steel until it is snug. I tie off the rebar about every 6 ft. to 8 ft. down the line.

Once the footing forms are complete and the steel is in place, make one last check of the elevations with a transit or a level. Adjust them accordingly, and then remind your crew not to use the forms as a convenient step for climbing out of the crawl space.

Concrete—Around here, I can call the ready-mix dispatcher a couple of hours before I need the concrete and give him a tentative delivery time and quantity. If I have questions about the mix, the dispatcher can usually answer them. Some suppliers even have a field representative who can drop by the job site to answer questions and help verify quantities. The dispatcher will ask you what strength concrete you want, and how much slump (a measure of the consistency of concrete) you want it to have. A typical footing mix should be rated at least at 2,500 psi, with a 5-in. to 7-in. slump.

Depending on what part of the country you're in, the dispatcher will also ask you if you want fly ash or water reducer in the concrete, or if you want it to be air entrained. Air-entrained concrete has tiny air bubbles in it to improve its workability during cold weather. Water reducers allow the concrete to flow more easily without reducing its strength.

Fly ash is a by-product of coal-burning power plants. When used in a concrete mix for footings and walls, it enhances the mix in many ways. It slows down the set, adds more fines to the mix (which makes for smooth walls), helps the concrete to pump easily and strengthens the mix. Best of all, fly ash reduces the overall cost of the concrete.

Fly ash can be used as either a substitute or an additive in a mix design. A 15% fly-ash substitute for portland cement in a five-sack mix (five sacks of cement per cubic yard of concrete) is most common for a 2,500 to 3,000-psi spec. Generally speaking, suppliers pay $70 per ton for portland and only $15 per ton for fly ash. Try it in your walls and footings if fly ash is available in your area. But be careful on flatwork because fly-ash concrete tends to be sticky and hard to finish with a trowel.

While you are waiting for the ready-mix truck to arrive, take a look around the site and make sure the truck can get to as many sides of the forms as possible. If some are inaccessible, make a chute out of lumber to extend the truck's chute. Where the two meet, prop up the ends with a sawhorse.

There is nothing quite like the calm before you hear the diesel working its way toward the job. Take that time to make some decisions about who is going to do what. You will need at least one helper, and a second won't hurt. One worker runs the chute in concert with the driver while others push, pull and poke the wet concrete into the forms. Shovels and lengths of 2x4 are the typical tools for these tasks. Pound the sides of the forms with a rubber mallet every

few feet to discourage any voids from forming in the concrete.

Concrete with a 5-in. slump does not "seek its own level," and it takes a lot of work to move it around the forms. The temptation to add water to the mix is a strong one, but remember that extra water weakens the concrete. According to research conducted by the Portland Cement Association, every gallon of water added to a yard of concrete reduces the strength of the mix by 5%. Some builders order stronger concrete (a six-sack mix instead of a five-sack mix) in anticipation of cutting its strength back with water so that they can move it around the forms more easily. Another way to make the concrete flow better is to order it with a water reducer.

Once the forms are filled with concrete (or filled to the chalkline), use a straight piece of wood to screed it flat. After screeding, use a wood float to level the top of the concrete. The wood float will also roughen the surface, which improves the bond between the footing and the stemwall to come.

While the concrete is still wet, you've got to make a shear connection for the stemwall. This can either be a groove in the top of the footing, a row of bricks set on edge or steel pins made of rebar. I prefer the pins because they are inexpensive and easy to install. I cut rebar into 4-ft. lengths and push them about 5 in. into the leveled footing on 4-ft. centers (photo facing page, top right). Without this precaution, it's conceivable that the backfill pressure could cause the stemwall to slide off the footing.

Form materials—A standard batch of concrete weighs about 140 lb. per cu. ft., and it exerts considerable pressure on the form work—especially at the bottom. To keep walls straight and true, you need a material that is able to resist that kind of pressure. To build 4-ft. stemwalls, I use ¾-in. CDX plywood reinforced with a 2x4 framework. When the foundation is complete, I disassemble the forms and recycle the plywood as subflooring and the 2x4s as wall framing and blocking.

At assembly time, it's best to have all the form materials stacked inside the footing perimeter. I build a jig to crosscut the studs to 45-in. lengths. This dimension allows room for the thickness of the top and bottom plates. Then I make as many 16-ft. panels as I can with continuous 16-ft. plates and two sheets of plywood. The 2x4 frames are built first, and then I nail down the plywood. I space the studs on 16-in. centers, although 24-in. centers would do if the concrete were placed slowly. I tried assembling 2x4 frames with a nail gun, but I found that the nails are very difficult to remove after I'm through with the form work. So I stick to 16d box nails for putting the frames together.

Before I nail the plywood to the frames, I use a jig to locate the ⅝-in. holes for the snap ties. They fall on 16-in. centers between the studs, forming two horizontal rows (photo facing page, bottom left). One row is 12 in. down from the top of the form; the other is 12 in. above the base of the form.

After drilling the holes, I set the plywood atop the frames and nail the plywood off with two or

Footing rebar can be propped up on rocks, bricks or commercial rebar stands called chairs, bolsters or dobies. Rockhill doesn't like to worry about the rebar slipping off its perch during a pour, so he wires them to the forms (above). Rebar pins set into the concrete on 4-ft. centers (right) will help the stemwall resist lateral pressure. When pushing them into place, don't let them go beyond the concrete and into the soil, where they can rust.

Exterior forms are braced with 2x4 tiebacks to keep them aligned as the concrete is placed (above). The short section of form in the foreground is used to complete the run of modular form panels. At outside corners (below), overlapping walers reinforce the forms.

Two rows of snap ties protrude from the exterior half of the stemwall forms. The ties make a handy place to secure the top and bottom courses of rebar. The middle piece is wired to the vertical rebar. With the steel in place, the forms are sprayed with a mixture of motor oil and diesel fuel to keep the concrete from adhering to the plywood.

Preparing for the pour. Wiring rebar in position is a breeze with twister ties and a winder (top left). Metal wedges driven onto the ends of the snap ties press the walers against the stud framework of the forms (top right). As the concrete is distributed around the forms, a worker follows along with the vibrator to eliminate any voids in the concrete (cover). A piece of plywood used as a deflection shield (above) directs the flow of concrete from the chute into the forms. Once the fresh concrete is screeded flat across the tops of the forms, anchor bolts are set so their tops extend about ½ in. beyond the thickness of the plate material (right). Anchor bolts should be at least 1 ft. from the end of a wall and are usually set 4 ft. o. c. The nut is threaded onto the bolt to keep the wet concrete from clogging the threads.

three 8d nails per stud. On most panels, I run the plywood flush with the edges of the perimeter 2x4s. For panels that form inside corners, I let the plywood run 4¼ in. past the outside stud so it can overlap the intersecting panel.

When most of the panels are ready and the footing formwork has been stripped, I snap a chalkline on the footing to mark the outside edge of the stemwall. I erect the exterior forms first, nailing the neighboring panels to one another through the butting studs. Then I brace the panels with stakes and tiebacks placed about 4 ft. o. c. (middle right photo, p. 23). Once I've got the big panels up, I make a custom infill panel to complete the exterior wall of the forms.

The snap ties are inserted when the exterior walls are up. Their stems are sandwiched by two horizontal 2x4 walers (drawing, p. 21). The assembly is then snugged together by driving home the metal wedges. It's okay to let the walers run past outside corners to avoid making long 2x4s into short ones, but be sure to offset butt joints in the walers to avoid weak spots.

Overlap the walers at outside corners and nail them together (bottom right photo, p. 23). As good as snap ties are, they won't hold the corners together. So the intersecting walers are essential for keeping the forms solid at corners.

When all the exterior forms are up, check them for level and square. Because subsequent work will be affected by the accuracy of the stemwalls, their alignment is critical. Check all 90° corners of the foundation using the 3-4-5 method. Level the top plates using a builder's level or a transit rather than a spirit level. Use shim shingles to make the adjustments.

I attach the bottom plates of the leveled and squared forms to the footing with powder-actuated 16d fasteners in every other bay. Then I use a garden sprayer to coat the inside surface of the forms with oil to make it easier to remove them from the concrete. I use 1 gal. of used engine oil with 4 gal. of diesel fuel as a spray mix.

Rebar and interior forms—A typical 4-ft. stemwall will have three pieces of #4 rebar running horizontally on 12-in. centers. One piece runs down the middle; the other two are 1 ft. in from the top and bottom. It's no accident that the snap ties are on the same levels as the top and bottom lengths of rebar because they support these pieces (tie wire holds them in place). The middle rebar is tied to the vertical rebar pins (bottom left photo, p. 23).

Whenever I have a lot of rebar to tie off, I make sure I've got a good supply of twister ties (photo facing page, top left) and a couple of winders. The ties have loops at both ends. The loops fit over the winder's hook, and with a few cranks on the handle the wire is twisted tight. They save a tremendous amount of time.

After the steel is set, check the forms for spots that still need spraying with oil. But be careful not to get any oil on the reinforcing steel, because it will interfere with the bond between the concrete and the steel. Spray the inside surface of the interior forms before you put them in place.

Lifting the interior forms into position differs slightly from raising the exterior forms. At inside corners, one of the forms has a plywood overlap, and you've got to thread the snap ties through the holes in the plywood as the panels are raised. Prop your 16-ft. panels on the footing edge and start to tilt them up until they meet the ends of the snap ties. Thread the lower course of ties through the holes first. After you have done the same with the top course, you have closed the form.

After all the interior walls are up, lock them together with the walers and wedges (photo facing page, top right). Work carefully—the cones will break if you pound the wedges down all the way. Snug is enough. Now use a spirit level to align the tops of the interior and exterior forms. When you're done, secure the bottom plate to the footing with powder-actuated fasteners.

When the forms are up, check the plans and locate blockouts for beam pockets, crawl-space vents, sewage and water-supply lines. You can make these out of dimension lumber, short pieces of plastic pipe or hunks of Styrofoam that have been glued together. For vents, I typically make a box out of pressure-treated 2x stock that has been ripped to a width of 8 in. I tack it to the inside of the forms, where it remains cast in the concrete as a nailer for screening and trim.

None of the walls in the project depicted here is longer than 32 ft., so control joints weren't required. Generally speaking, walls that are longer than 36 ft. require a break in the wall to control cracking. Control-joint theory and practice are beyond the scope of this article. But if you need to know more about them, the Portland Cement Association (5420 Old Orchard Road, Skokie, Ill. 60077) publishes an informative pamphlet on their function and installation.

The pour—A 4-ft. stemwall, 8 in. wide and 10 ft. long, will contain about 1 cu. yd. of concrete. I calculate the amount of concrete I need to fill the forms, and I round up to the nearest half-yard to be on the safe side. Now I am ready to place my order.

Most builders use a five-sack mix for stemwalls, which yields 2,500 to 3,000-psi concrete. I like my walls to be smooth, so I prefer to use a six-sack mix (3,500 to 4,000 psi) and add a little water to it to get a 6-in. to 7-in. slump. I add the extra portland to the mix to offset the loss of strength caused by the additional water.

As the weight of fresh concrete in the forms doubles, the stress on the bottom of the forms quadruples. For this reason, it's a good idea to fill the forms in stages, or lifts. I do 4-ft. walls in two lifts. By the time I get all the way around the perimeter, the concrete has had a chance to take its initial set, taking some of the burden off the forms.

A good driver can regulate the flow off the chute and move the truck ahead slowly to make the pour a pretty slick operation. A plywood sheet about 2 ft. square is handy for diverting the concrete into the form as it comes off the chute (middle photo, facing page). Before the truck arrives, get a concrete vibrator or a rubber hammer, a wooden float and your anchor bolts.

As one worker directs the flow of the concrete into the forms, another should follow, stinging the concrete with a vibrator or rapping the forms with a rubber mallet. This is essential for filling air pockets and eliminating segregation. It also stresses your formwork to the maximum. You can feel it in your feet as the entire assembly resonates from the vibration. Overvibration of the concrete is a common mistake—three to five-second spurts every 3 ft. is plenty.

When you make the second lift be sure to vibrate the joint between the two. Sting the concrete a little extra around blockouts to eliminate the air pockets that typically occur there. As the concrete reaches the top of the form, strike it off level using the edges of the form as a screed. Don't forget to start setting anchor bolts before the concrete sets (bottom photo, facing page).

After the bolts are placed, scrape off any concrete that got on the forms and go home. Let everything sit for at least 24 hours (48 hours is better). I always try to get the concrete in before the weekend so we can start stripping the forms on Monday. It usually takes a day to strip all the forms, pull all the nails and stack the lumber.

I spray the green concrete with a curing compound, which slows the loss of water from the concrete. Both resin and latex-base curing compounds are available. I prefer the latex because it is less likely to clog the sprayer. If you plan to top the concrete with another finish, remove the curing compound with a sandblaster. Curing compound smells terrible but is worth the money because concrete has to cure properly to achieve its rated strength. Ignoring this step can reduce the concrete strength by as much as 50%. Remember, good concrete costs money and bad concrete costs more.

Gravel, insulation and backfill—I like to get gravel into the crawl space as soon as we get the form wood out. I use the smallest size I can afford, keeping in mind the hours someone will spend kneeling on it running mechanicals. Small gravel is comfortable, big is torture.

I also run insulation inside the walls. For a small foundation like this one, I put 4 in. to 6 in. of beadboard on the inside. This insulates the crawl space quite well.

If you backfill too soon, you will surely crack the concrete. After a few days, the concrete will be strong enough to take the compression loads of floor framing, but not the force of an entire truckload of gravel hitting it from the side. I wait to backfill at least until I've got my first-floor wall framing up. Concrete achieves most of its strength in 28 days.

Further information on concrete systems can be obtained from the Portland Cement Association (5420 Old Orchard Road, Skokie, Ill. 60077). Write for their free catalog of brochures. The National Association of Home Builders (15th and M Sts., N.W., Washington, D. C. 20005) has a useful publication entitled "Residential Concrete" ($15 to non-members, $12 to members), and the American Concrete Institute (Box 19150, Redford Station, Detroit, Mich. 48219) also has numerous publications available on the subject. ☐

Dan Rockhill is a contractor and an associate professor of architecture at the University of Kansas at Lawrence. Photos by the author.

Pouring Concrete Slabs

Tips on ordering, placing, screeding, floating and finishing

by Carl Hagstrom

Using a screed rail. Two workers use a magnesium straightedge, or screed rail, to level freshly placed concrete. The ends of the rail glide over previously leveled concrete strips, called wet screeds. A third worker rakes the concrete behind the screed rail to adjust for high and low spots.

L iquid stone. It's an image you might think describes placing concrete, and to some extent, it does. But there's more to it than backing up the ready-mix truck, opening the spigot and letting the concrete flow out until the forms are full.

The applications of concrete are almost limitless, but here I'll focus on residential slabs. About half of the new homes currently built in the United States start with full-basement foundations, and virtually all of these basements have concrete floors. For the most part, these floors consist of 4 in. of concrete placed over 4 in. or more of crushed stone. Concrete floors in garages are similar, except they are sometimes reinforced with wire mesh or steel. But whether it's a basement or a garage slab, the way you place the concrete is the same.

Ordering concrete—Concrete is sold by the cubic yard, and calculating the amount you need is simple: length times width times depth (in feet) divided by 27 equals cubic yards. Most concrete trucks max out at 9 yd., and if your floor will require more than nine (the average floor uses about 19 yd.), tell your supplier to allow about an hour to an hour and a half per truckload so that all the trucks don't arrive at once.

But a word of caution. Running out of concrete is like running out of champagne at a wedding: If you can't get more real soon, you're headed for trouble. Don't be stingy with your concrete estimate. You're a lot better off with half a yard left over than a quarter yard short.

Once you've told your concrete supplier how much concrete, you'll have to tell them what kind. Concrete is made up of four basic ingredients: cement, sand, stone and water. Depending on the proportions of the ingredients, the strength can vary considerably. Compressive strength, measured in pounds per square inch (psi), is the method used to evaluate the performance of a given mix. Generally speaking, the higher the cement content, the higher the compressive strength. Most residential concrete has a compressive strength between 2,000 psi and 3,500 psi.

You'll also need to specify the slump, or the wetness, of the mix. A slump of 4 to 5 is about right for slabs, whereas a slump of about 2 to 3 is normal for piers, which don't need to be worked, so the concrete can be stiffer.

Be prepared—Take a few moments and survey the situation. Do you have a grade-level door, or will you need to chute the concrete through a basement window? Will the ready-mix truck be able to get next to the house, and if not, will the manpower be available to transport the concrete in wheelbarrows? Pushing one wheelbarrow full of concrete uphill is possible for some, but making 30 trips uphill is a job for the John Henry type. It never hurts to have more help than you might need because concrete is always a rugged day's work.

A little rain the night before can turn a dry approach into a muddy nightmare. I call my supplier several days in advance and say that I'm shooting for next Thursday, for example, but that

First mud. With the vapor barrier in place and the chalklines snapped, the first load of concrete is dumped in the far corner of the foundation. The mason dumps the concrete away from the wall so that he won't cover the chalkline.

Establishing a perimeter screed. A magnesium hand float is used to push the concrete up to the chalkline. This strip of wet concrete, placed along the foundation walls, is a perimeter screed.

Raker's role. As the screed rail levels concrete between the perimeter screed (right) and the center screed (left), the raker pulls away excess concrete or fills low spots. A rebar spike set at finish-floor height and subsequently driven below the concrete's surface establishes the center screed's level.

I'll call first thing Thursday to confirm. If conditions are terrible, I reschedule.

Remember, concrete waits for no one. From the minute it leaves the plant, it has a finite time before it sets up, and just about any builder can come up with a horror story describing a pour that got away.

Placing the slab—Arrive early on the day of the pour and use a water level or a transit to snap chalklines on the foundation wall at finish-floor height (usually 4 in. higher than the stone). The lines help you level the concrete along the walls.

You should also lay out the vapor barrier at this time. Six-mil polyethylene works well, but if you're concerned about punctures from traffic during the pour, a puncture-resistant, cross-laminated product is available, called Tu-Tuf (Sto-Cote Products, Inc., P. O. Box 310, Richmond, Ill. 60071; 815-675-2358).

If you elect to use wire-mesh reinforcement, this is also the time to lay it out. Wire mesh doesn't prevent cracking, but it will help keep hairline cracks tight, even as the temperature varies. Typically, a basement slab isn't subjected to wide temperature swings. Therefore, a basement slab placed over a properly prepared stone

base doesn't require wire mesh. Garage slabs, on the other hand, typically experience harsher weather conditions, and wire mesh may be used as temperature reinforcement. But wire mesh won't be effective unless it's placed midway in the thickness of the slab, so be sure to use wire high chairs, which hold the reinforcement up off the stone during the pour.

The ready-mix truck arrives, and the driver asks, "How wet do you want it, Mac?" Drivers routinely ask about adding water to soften the mix. When a mix is too stiff, it's physically difficult to work and presents problems when it's time to float and finish the slab. To get a smooth, hard, dense finish on top of the slab, the mix has to be workable. As mentioned earlier, however, the wetness was determined when you specified the slump of the mix. And as any structural engineer will tell you, when you add water to concrete, you lower the final strength. The issue of water content in concrete is critical; many concrete companies require that you "sign off" on the delivery slip when requesting additional water so that they have a record of your compromising the rated strength of the mix.

If the first few wheelbarrows of concrete are difficult to work, have the driver add water to it—

but in small amounts. You can always soften the mix by adding water, but you can never dry it if it becomes too wet.

Leveling with wet screeds—There are many ways to place a basement slab. If you've never placed one, ask some masons about the techniques they use. If you have placed a few slabs, don't be afraid to try a different method; you may discover a system that you're more comfortable with. But whatever approach you take, follow a logical progression: Don't trap yourself in a corner. I prefer to use wet screeds as guides to level the slab (photo p. 26).

Wet screeds are wet strips of concrete that are leveled off at finish-floor height and used to guide a straightedge, or screed rail, as you level the slab. If you've ever watched a sidewalk being placed, you've seen concrete placed between two wood forms, a screed rail placed on top of those forms and sawed back and forth to strike the wet concrete down to the level of the forms. Wet screeds guide the screed rail in places where there are no wood forms, such as against an existing concrete wall or in the middle of a slab.

Where you start with your wet screeds depends on the layout of the slab. In a typical rectangular

Operating a bull float. After the concrete is screeded, a bull float pushes stones down and brings up the fines—sand and cement—that make a smooth, finished surface. To use a bull float, lower its handle as you push it away, and lift the handle as you pull it back.

basement with the walls already in place, a wet screed is placed around the perimeter of the foundation, and a second wet screed is placed down the center of the foundation (photo facing page), parallel to the longer dimension of the foundation. On a bigger slab you might need more wet screeds; the determining factor is the length of the screed rail you'll be using.

Placing the wet screeds around the perimeter of the foundation is simple. Use the chalkline you snapped at finish-floor height as a guide to level the concrete at the wall (bottom photo, p. 27). As the concrete is placed, either from a wheelbarrow or directly from the chute, use a magnesium hand float to push and level the concrete to the line. Be sure you don't cover up your chalkline as you place the concrete. Dump it near the wall and bring it up to the line with the float (top photo, p. 27).

Establishing the level of the center screed requires that you drive pins about 8 ft. apart at the level of the finish floor; 16-in. lengths of ½-in. rebar work well. Try to set these pins immediately before the pour, using a transit or a stringline, and cover them with upturned buckets so that no one trips on the pins. Place and level a pad of concrete around each pin, then fill in the area between the pads with concrete and use a screed rail, guided by the pads, to level the area between them. As you complete each portion of this center screed, drive the pins a few inches below the surface with a hammer and fill the resulting holes with a little concrete.

Raking and striking—To fill in the areas between screeds, place and rake the concrete as close as possible to finish level before striking with a screed rail. Placing too much material makes it difficult to pull the excess concrete with the screed rail as you strike off, and the weight of the excess concrete can distort a wooden screed rail. If you starve the area between the screeds, you'll constantly be backtracking through freshly placed concrete, filling in low spots and rescreeding.

Cement shoes. Kneeboards—pieces of plywood with strips of wood on one edge—allow you to move around on fresh concrete without sinking in because the boards distribute weight over a large surface area. Concrete finishing is done from kneeboards while the concrete is in a plastic state, meaning it's neither liquid nor solid.

Using a screed rail, strike off the concrete with the perimeter screed and the center screed as guides. Your path of escape will determine the placement and the size of your screeds, but generally speaking, you progress in about 10-ft. or 12-ft. sections of slab.

The person raking the concrete can make or break the pour. As the wheelbarrows are dumped, the raker should nudge the concrete to the plane of the finish floor, eyeing the placed concrete like a golfer lining up a putt, and noting any mounds or valleys that will create problems as the screed rail works across. As the concrete is struck off, an alert rake person will pull away any excess concrete accumulating ahead of the screed rail (photo facing page) and push concrete into any low spots.

At this stage of the pour, with five or more people working, teamwork is the name of the game. Establish each person's role well ahead of the pour, and do your best to stick to the plan.

Striking off the concrete with the screed rail is the last step in placing the concrete and the first step in finishing it. A good, straight 2x4 will work well, but magnesium screed rails, available in various lengths, will perform better. No matter which you choose, working a straightedge back and forth is a lot like running a two-man saw. The work is done on the pull stroke, and you have to be aware of your partner's progress. Wear rubber boots because, standing alongside the wet screeds, you'll often be wading through concrete. (For anyone tempted by the prospect of a barefoot frolic in concrete, be warned that concrete is caustic and will corrode your skin.)

As you saw the screed rail back and forth, let it float on top of the wet screeds, keeping an eye open for low spots and stopping when excess concrete dams up ahead of the screed rail so that the raker can pull off the excess.

Bull floating—As the pour progresses, it's necessary to smooth the surface of the leveled concrete with a magnesium bull float (top photo, this page). When you bull float is determined by

Start with a mag; finish with steel. After bull floating, use a magnesium float (left) to smooth out bumps and fill in low spots. The resulting finish will be coarse. Later, use a steel trowel (right) to get a smooth, dense finish that won't crumble when it's swept.

Two hands! Two hands! As the concrete sets up, working it with a steel trowel may require the strength of two hands. The back of the trowel is angled up as you push it away (left), and the front of the trowel is angled up as you pull it toward you (right).

the length of the tool's handle and how comfortable you are operating the tool. For example, a bull float with an 18-ft. handle will easily float a 10-ft. or 12-ft. section of a slab. Bull floating levels the ridges created by the screed rail, but more importantly, it brings cement and sand to the surface of the slab and pushes stones lower.

Water is the lightest ingredient in concrete and quickly finds its way to the surface as you jostle the mix around with a bull float. As the water rises to the surface, it also brings some cement and sand with it. These are the fines (sometimes called fat or cream) that provide a stone-free medium for troweling to a smooth finish.

Although its size is intimidating, a bull float works about the same as a hand trowel. The trick is to keep the leading edge of the bull float inclined above the surface of the slab by lowering

the bull float's handle as you push it away and raising the handle as you pull it back. Some masons jiggle the handle as they move it out and back to jostle more fines to the surface. The ease of final troweling depends on how well the slab has been bull floated.

Hurry up and wait—Once the slab is placed and bull floated, it's time to sit and wait. The first stage will be the evaporation of the bleed water, water that rises to the surface as the slab sets up. Depending on the weather conditions and the consistency of the mix, this time can vary from one hour on a hot, dry day to 10 hours on a cool, damp day (see sidebar on the facing page).

But keep in mind that when concrete starts to set, it waits for no one. There is a small window of opportunity in which you can work the slab, and

if you happen to be out for coffee when the concrete starts to set up, you'll learn an expensive lesson. Unless you're a veteran finisher, don't ever leave the pour; you may return to a problem whose only solution is a jackhammer.

Floating from kneeboards—Once the bleed water has evaporated, work the slab. Some slabs (in crawlspaces, for example) are acceptable with just a coarse, bull-floated finish. But these finishes tend to dust over time; that is, concrete particles come loose from the coarse slab surface whenever it's swept. Additional finishing compacts the surface so that the slab won't dust.

You may have seen professionals using a power trowel to float and finish larger slabs. A power trowel works like a lawn mower without wheels. It rides on rotating blades that smooth the surface of the concrete. (For more on power troweling, see p. 38.) If you're inexperienced, however, or if the slab is small, you're better off finishing it by hand. And even professionals still use hand floats and trowels at the edges of the slab and around projections because a power trowel will only finish to within a few inches of these spots.

Hand finishing is commonly done from kneeboards, which are like snow shoes for the still-wet slab (bottom photo, p. 29). They let you move around on the slab without sinking. To make a simple pair of kneeboards, cut two pieces of ¾-in. plywood 2 ft. square, and tack a 2x2 strip at one edge of each piece.

It's difficult to describe just when the slab is ready for hand floating, but it may help to think of the slab as drying from the bottom up. If you set a kneeboard on the slab, and it sinks ¾ in. when you step on it, you're too early; if it fails to leave a mark, you're too late.

As soon as you can easily smooth over the tracks the kneeboards leave behind, the slab is ready for the first hand floating.

Test the concrete at the area where the pour started because it tends to be ready first. If any areas of the slab are in direct sunlight, you can bet they'll be ready long before the shaded areas are. At any rate, your first pass will be with a magnesium hand float (top left photo, facing page).

Like a bull float, a magnesium hand float works the fines to the surface, and you fill in any low spots or knock down any high spots during this pass. The goal when using the magnesium float is to level the concrete, preparing a surface that is ready for smoothing with the steel trowel. You can generally work the entire slab with the magnesium float before it's time to trowel with steel.

The difference between a magnesium float and a steel trowel is easy to recognize on the slab. You can work the slab all day long with magnesium, but you'll never get beyond a level, grainy surface. But when the slab is ready, and you lay a steel trowel to it, the results are impressive.

Hit the slab with steel—Keeping the image in your mind of the slab drying from the bottom up, picture the top ⅛ in. of the concrete, which is all cement, sand and water. While this top section is in a plastic state—neither liquid nor solid—the steel trowel will smooth this layer and compact it into a dense, hard finish. Now the preparatory work pays off; if the concrete was placed and leveled accurately, the final finish goes quickly.

Obtaining an exceptionally smooth finish is a practiced technique that takes years to develop. The steeper the angle of the trowel to the slab, the more trowel marks will occur. If you hold the trowel at an extremely slight angle, you're liable to catch the slab and tear out the surface.

Your troweling technique will be dictated by how loose or tight the surface of the slab is. When the surface is wet, you can hold your trowel fairly flat, but as the fines tighten up, you'll have to increase both the angle and pressure of your trowel. As the slab dries you might have to use both hands on the trowel to muscle some fines to the surface (bottom photos, facing page). Once the fines have emerged, switch back to one hand and polish the area with your trowel (top

Tips for pouring in the weather

Temperamental is a literal description of concrete. Temperature, along with humidity, influences the pour more than any other factor.

Hot-weather pours—When it's hot, and the humidity is low, every minute is important. If you spend time fussing around, when the last wheelbarrow of concrete is finally off the truck, the first section of floor you placed will probably be hard enough to walk on.

Here are some strategies that help in hot weather:

Even if a polyethylene vapor barrier is not required, use one. It blocks the moisture from dropping through the subgravel.

Have lots of help available. The sooner you get the truck unloaded and the concrete leveled, the better your chances will be of getting a good finish.

Have two finishers working the slab: one with a magnesium float, and another following behind with a steel trowel.

Although it compromises compressive strength, consider using a wetter mix to buy a little more working time.

If more than one truckload is needed, coordinate the arrival times carefully. If a fresh truckload of concrete has to sit and wait an hour while you finish unloading the first truck, you may find that concrete from the second truckload will set up before you're ready for it.

Areas that receive direct sunlight set up much quicker than shaded areas.

Start wetting down the slab as soon as the final finish has set. Few things will weaken concrete as much as a "flash" set, where the concrete dries too quickly.

Cool-weather pours—When the temperature is cool, concrete initially reacts in slow motion. After the slab is placed, and the bleed water slowly evaporates, you'll wait hours for the slab to tighten up enough to start hand troweling. When it's finally

ready to be troweled, you'd better be there because that window of opportunity for finishing doesn't stay open much longer on a cool day than it does on a warm day. Here are a few cool-weather tips:

Don't wet the mix any more than necessary.

If a polyethylene vapor barrier isn't required, don't use one. Any moisture that drains out of the slab will speed the set.

Pour as early as possible to avoid finishing the slab after dark.

Cold-weather pours—When the temperature is cold, a whole new set of rules comes into play. Concrete cannot be allowed to freeze. That tender, finely finished surface you just troweled on the slab will turn to mush if it's allowed to freeze. Fortunately, the chemical reaction that takes place when concrete hardens generates heat.

Here are some strategies that help in a cold-weather pour:

Ask your concrete supplier about using warm mixing water to prevent problems during transit on days when the temperature is well below freezing.

Having the supplier add calcium to the mix accelerates the initial set of the concrete, and the concrete achieves the strength to resist freeze/thaw stress faster. The amount of calcium is measured as a percentage of the cement content and ranges from ½% to 2%. Talk to a veteran concrete finisher before deciding when and how much calcium to add to the mix. Too much calcium produces the same problems as hot, dry weather. It's important to note that calcium is corrosive to steel and should never be used in steel-reinforced concrete.

Always be sure that all components of the subbase are frost free.

Provide supplemental heat to keep the building above freezing.

Cover the slab with polyethylene and then spread an insulating layer of straw or hay at least 4 in. thick on top, or use an insulating tarp.

The best strategy: Pour when cold temperatures are not an issue. —C. H.

right photo, facing page). If you've waited too long, and you're losing the slab, sprinkle water on its surface to buy a little more finishing time. After that, there isn't enough angle, pressure or water anywhere on earth to bring a lost slab back to life. If it's important that the final finish be first rate, consider hiring a professional. Remember, you get only one try.

Curing the finished slab—While it's true that you can walk on the floor the day after it's placed, concrete actually hardens very slowly. The initial set represents about a quarter of the total strength; it takes about a month for concrete to cure fully. The goal during this period is to have the concrete cure as slowly as possible.

Keeping the slab soaked with water for four or five days will keep it from drying too quickly, but

continual hosing down involves a lot of time and effort. Slabs can require a soaking every half hour in the heat of summer. A masonry sealer applied the day after the pour will keep the slab from drying too quickly and protect the floor from stains that might otherwise wick into the slab.

When you consider that the material cost of a basement slab is less than $1 per sq. ft., it's difficult to imagine a more economical finished floor system. But when you consider the cost of removing and replacing an improperly finished concrete floor, the importance of knowing how to handle two or three truckloads of concrete becomes apparent. □

Carl Hagstrom manages Hagstrom Contracting, a design/build company in Montrose, Pa. Photos by Rich Ziegner.

Avoiding Common Mistakes in Concrete and Masonry

An engineer offers a few simple changes to make walls and foundations much sturdier

by Thor Matteson

At the engineering firm where I work, the past few years have brought us about a dozen jobs retrofitting designs for relatively new buildings that were structurally deficient or failing for one reason or another. Typical was the work we did on a poorly designed office building. Improperly placed rebar substantially reduced the strength of a critical grade beam. After a good deal of excavation, we epoxied dowels into the old grade beam and reinforced it with 10 yd. of new concrete.

Now I notice structural problems everywhere I look. I can't even go to the supermarket without wincing ever so slightly at the shrinkage cracks in its concrete-block walls (photo right). Although these problems are how I make a living, many of them could have been avoided.

Concrete strength relies on the right mix and reinforcement—Reinforced concrete is barely a hundred years old, and engineers are still refining their assumptions of its properties. Yet some contractors (even large governmental agencies) have not changed their methods in the past 30 or 40 years.

Concrete alone lacks appreciable tensile strength. Steel reinforcing, or rebar, used in concrete cannot withstand compressive force by itself. Combining the strengths of concrete and steel produces the required structural properties (sidebar p. 37).

Anyone who works with concrete knows that steel reinforcement provides tensile strength. But even experienced builders and designers commonly overlook an obvious consequence of this fact. Rebar must extend into concrete deep enough to develop that tensile strength. Instead of pulling out of the concrete, the bars will start to stretch.

Problems commonly arise at points where rebar changes direction and at intersections. Take, for example, a corner in a footing that has two horizontal rebars. Workers often place the outer

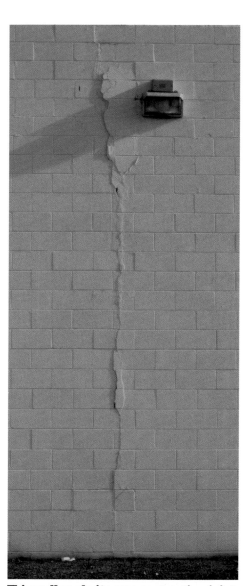

This wall made its own contraction joint. Quarter-inch contraction joints made of a high-grade elastomeric sealant allow a block wall to shrink as water dissipates from the grout, instead of cracking. This wall should have had these control joints every 15 ft. to 20 ft.

bar wrapping around the outside of the corner, which is correct. But then they wrap the inside bar around the inside of the corner, losing a few essential inches of development length (sidebar p. 37). Inside bars at corners should cross, run past each other and extend toward the far side of the footing (drawing facing page).

Rebar has to go deep enough to do the job—Straight reinforcing bars can develop sufficient bond if they extend far enough into the concrete. When you don't have thick enough concrete (such as at a wall corner or at a T-intersection), a hook at the bar's end may substitute for the lack of available embedment. If a perpendicular bar is placed inside the hook, it spreads the force to a greater area of concrete. (Usually there is a whole row of hooked bars, and a single bar can run through all of the hooks.) We like to see a bar inside the bend at any change in bar direction.

Sometimes the tail of a hook may not fit where it's shown on the plans. Usually you can rotate the tail of the hook to clear obstacles. As long as the hook extends into concrete to develop sufficient anchorage, it's doing its job. You should check with the designer first, though. Sometimes hook tails need to lap with other bars.

Less water in concrete means less shrinkage and more strength—Nuisance cracking in concrete has two major causes: shrinkage during the curing process and thermal expansion or contraction due to temperature swings. Generally, shrinkage is caused as the water in the concrete gradually dissipates. As much as one-third of the water, or 5% of the total volume of the concrete, can dissipate.

The strongest concrete uses only enough water to hydrate all of the cement in the mix. Excess water leaves space in the concrete when it evaporates, making the concrete less dense and therefore weaker. (Lost water affects the concrete's structure on a molecular level, unlike the tiny bubbles that are left in air-entrained concrete.) But strong concrete is worthless if you can't place it, and getting good workability requires more water.

Good workability means that you can consolidate the concrete around reinforcement and into corners by vibration. It does not mean that the concrete flows there by itself. If I overhear concrete workers complaining about how hard it is to work the mix, I know it's good concrete.

Reducing the amount of water in the mix also makes the cured concrete less permeable to air, water and salts. Although air seems harmless, the carbon dioxide it carries can react with water in a process called carbonation. In carbonation, water and carbon dioxide combine to form carbonic acid, a weak acid. Over time, enough acid lowers the concrete's alkalinity to the point where steel corrodes much more readily. Using a drier mix and consolidating it well can forestall carbonation for a long time.

Benefits of reducing the proportion of sand in concrete—Using a high proportion of sand makes concrete easier to finish, but it causes problems, including increased shrinkage and reduced strength. A mix with as little sand as possible helps ease these problems.

Reducing the amount of sand makes concrete harder to work, but it also means fewer shrinkage cracks. Here's why: To make concrete workable, water must coat the surfaces of all of the aggregate. For a given weight of aggregate, large pieces will have considerably less surface area than small grains, and thus require less water to provide lubrication. Substituting pea gravel for some of the sand in the mix allows you to reduce the amount of water needed for workability. Less water is left free to evaporate, and the concrete will crack less.

When you order a load of concrete, you can specify less sand and more gravel. Your local batch plant should have various mix designs on file so that you can specify a 60-40 or 65-35 mix (the ratio of gravel to sand).

Less water also reduces potential reactive aggregate problems. Some aggregates will gradually react with the alkaline cement and expand slightly in the process. When this situation happens, the concrete's own ingredients break it apart, and it gradually disintegrates. Although coarse reactive aggregates can be sorted out economically, a similar process does not exist for sand, so the less sand the better.

Special considerations for slabs on grade—Concrete slabs need to have a firm, even, well-compacted substrate, which begins with properly prepared original ground. Remove all sod, stumps, roots, other organic matter, large rocks

Rebar must be detailed carefully at corners

At intersections and corners of concrete footings and walls, problems can arise if reinforcement is improperly placed. Steel bars must overlap the correct length and should hook around perpendicular reinforcing. At corners, inside bars should cross, run past each other and extend to the far side of the footing. Otherwise, the inside bar in the corner lacks sufficient embedment, and the concrete then may pop out. The three sets of drawings below are plan views of footings.

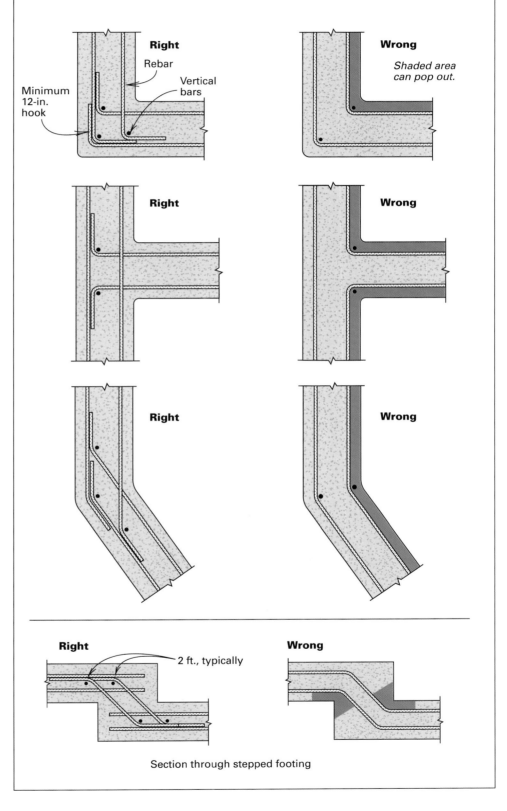

Section through stepped footing

and wet, mushy soil. The soil must be compacted thoroughly and evenly. If you have unstable soil, plan on hiring an engineer or ending up with a cracked, buckled slab.

Placing gravel, crushed rock or sand before pouring the slab makes an even surface to receive the concrete. As the concrete shrinks, the slab edges want to slide toward its center. A smooth surface of sand makes it easier for the concrete to slide, reducing its tendency to crack.

A layer of sand or gravel also creates a capillary break between the ground and the slab. Sand still wicks some water, so we usually recommend placing a vapor barrier on top of it. You should compact any subbase material well and dampen any exposed areas just before pouring the slab.

Rebar performs better than wire mesh in slabs—Steel rebar and wire mesh both serve the same function: They add structural reinforcement to the concrete. Rebar is only slightly more expensive than mesh but easier to keep centered in the slab. If mesh is properly placed, it should do just as good a job of strengthening the concrete as rebar.

Reinforcement should stay centered in the slab and not trip concrete workers as they move about. However, if the reinforcement is wire mesh, there's no way to avoid trampling it while placing the concrete. So the mesh gets embedded in the sand beneath the slab. Although it's not impossible to keep the mesh centered uniformly in the slab (photo left), it is difficult.

Placing a grid of #3 rebar at 18-in. centers provides adequate slab reinforcement and places for nimble feet to step. Supporting the bars on metal "high chairs" or precast concrete cubes ("dobies" where I live) keeps them in the proper position as concrete is poured over them. The cost difference between mesh and rebar is almost negligible, and the reinforcement ends up where it belongs (drawing left; photo left).

Reinforcement falls into two main categories: structural and shrinkage/temperature. Structural reinforcement provides strength to resist bending, compression or tensile loads. Shrinkage/temperature reinforcement reduces the concrete's tendency to crack as it dries or as it contracts or expands due to temperature changes.

For the latter, chopped fibers of polypropylene, nylon, steel, glass, palm fronds or the like may be added to the concrete mixture. Because the fibers are automatically distributed throughout the concrete during mixing, there are no concerns about proper placement. For most slabs on grade that require no structural reinforcement, fiber reinforcing can be placed in the mix at the batch plant.

Control joints help control cracking—Shrinkage cracks appear in any slab. But you can keep them small and govern where they appear by building in control joints. A control joint works like perforations in a piece of paper. The joint is a line of weakness in the slab that eventually becomes a crack.

If you place the joints at 15-ft. to 20-ft. intervals, most cracking occurs along the joints. For large industrial or commercial slabs, control joints usually are sawn into the concrete. On smaller jobs it may not be worthwhile to bring a concrete saw on site. In such cases, long pieces of plastic extruded in T-shaped cross section can form the joints. Concrete finishers force the stem of the T into the slab. To ensure that cracks occur at the control joints, their depth should be at least one-quarter of the slab's thickness.

A few tips on placing and working concrete—Concrete-form construction is beyond the scope of this article. But when you build forms, build them stronger than necessary. Brace them well; expect to climb all over them while carrying an ornery vibrator.

Secure anchor bolts and other inserts to the forms in their proper locations before the pour.

Wire mesh can end up at the bottom of the slab. This core sample of a wire-mesh reinforced concrete slab illustrates what happens when concrete is poured over the mesh. Rather than staying in the middle of the slab, where it strengthens the slab, the mesh gets pushed to the bottom, where it does little good.

Rebar on chairs stays in center of slab

In concrete slabs, steel-bar reinforcement works better than wire-mesh reinforcement because it stays where it's put. It's also important to have a firm, even bed beneath the slab and to cover the substrate with a plastic vapor barrier.

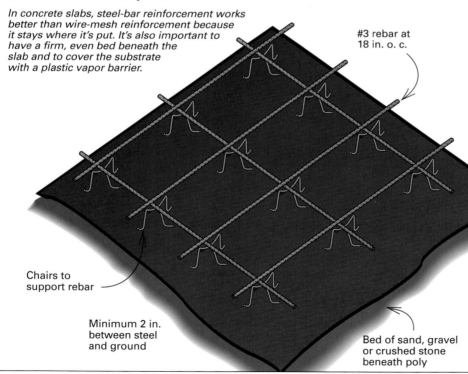

#3 rebar at 18 in. o. c.

Chairs to support rebar

Minimum 2 in. between steel and ground

Bed of sand, gravel or crushed stone beneath poly

(Poking bolts into wet concrete disturbs the aggregate and gives a weaker bond than pouring the mix around the bolt.) If you use a form-release agent, apply it to the form boards only, not to the reinforcement.

Always place concrete as close to its final position as possible. When concrete comes out of the chute or the pump hose, it should not free-fall more than 4 ft., or clatter off the rebar or forms (drawing below). Either of these conditions can cause the coarse aggregate to separate from the rest of the mix, resulting in concrete that's not uniform. Preventing this condition in tall walls or columns usually requires that a concrete pumper dispense the concrete from a hose, which can maneuver close to the bottom of the forms.

The Uniform Building Code requires vibrating all structural concrete to eliminate voids. To do this procedure properly, turn the vibrator on and insert it into the concrete as quickly as possible. To vibrate the first lift poured, insert the vibrator all the way to the bottom for 10 seconds. Then withdraw it slowly, about 3 in. per second. The goal is to allow all of the trapped air bubbles to get out of the concrete. The bubbles move up slowly; if the vibrator head moves faster than they do, they will remain trapped. Vibrate any additional lifts the same way, but extend the end of the vibrator about 6 in. into the previous lift.

The vibrator influences a circular area of concrete, whose size depends on the power of the tool. These circles of influence should overlap. Use a regular pattern and consolidate the concrete. Do not insert the vibrator at haphazard angles or use it to move concrete in the forms.

Keep the concrete wet and warm—Builders always want to hurry to strip the forms, but leaving them in place a few days holds the moisture in the concrete. A week of wet curing would make any engineer happy. Keep the concrete wet by covering it with plastic or wet burlap, or with spray from a fog nozzle. Slabs also can be flooded. Curing compounds that seal in moisture when sprayed on concrete are available.

Rapid drying or freezing severely reduces concrete's strength and results in weak slab surfaces. A little time invested in proper curing protects the finished product you worked hard for.

For further information, the American Concrete Institute (810-848-3700) publishes several references and the model code, ACI 318-89. The Aberdeen Group (708-543-0870) offers several publications and references intended for contractors and builders. "Design and Control of Concrete Mixtures" by the Portland Cement Association (708-966-6200) offers much information. At $35 for 200 pages, it addresses mixing, placing, finishing, testing and more.

Understanding the principles of concrete-block construction—Reinforced concrete-block construction can produce strong, durable walls efficiently. But too few masons really understand the principles involved in the trade. In this type of construction, three components form the structural system: the blocks and the mortar that holds them together, the reinforcement and the grout, which is used to fill in the cores in the concrete block.

Grout is a mix of fine gravel, sand, cement and water. In some areas (not California), mortar is used instead of grout in block cores. However, grout contains coarser aggregate than mortar, which makes it stronger. The UBC requires each cell that contains reinforcement to be filled with grout. In seismic zones 3 and 4 (almost all of California and the West Coast), this rule means filling at least every sixth vertical cell (every 4 ft.) and grouting a bond beam at the bottom, middle and top of an 8-ft. wall (drawing p. 84). (Bond-beam blocks are made with space for horizontal rebar to lay in them; a course of these blocks, when reinforced with steel and filled with grout, forms a bond beam.) Additional bars at openings, or where required by the designer, often decrease the spacing to 32 in. or even 24 in. Grouting all of the cells is usually easier than trying to block off the cells where grout is not required, so most walls we see are solid-grouted. In essence, concrete blocks or withes of brick just serve as the forms for a fine-aggregate concrete (grout) wall. Your wall should solidly attach to the footing and act as a monolithic unit, not a stack of separate blocks. Also,

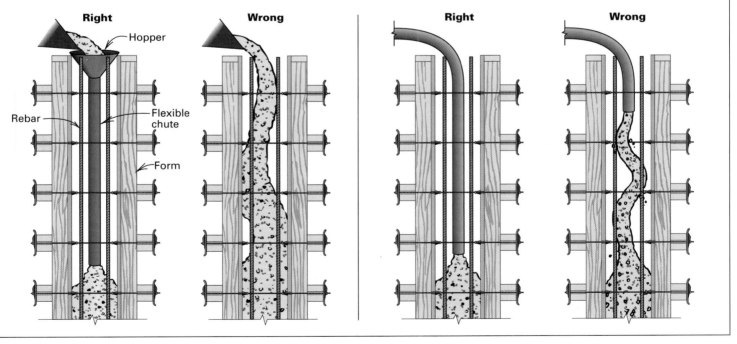

The right way and the wrong way to place concrete

If concrete is dropped from a height greater than 4 ft. into a form or permitted to fall freely over reinforcement, the aggregate can separate from the concrete or honeycomb at the bottom. When filling forms from a chute, use a hopper to deliver concrete to the bottom of the forms. When using a pump, feed the hose to the bottom of the form.

Right — Hopper

Rebar — Flexible chute

Form

Wrong

Right

Wrong

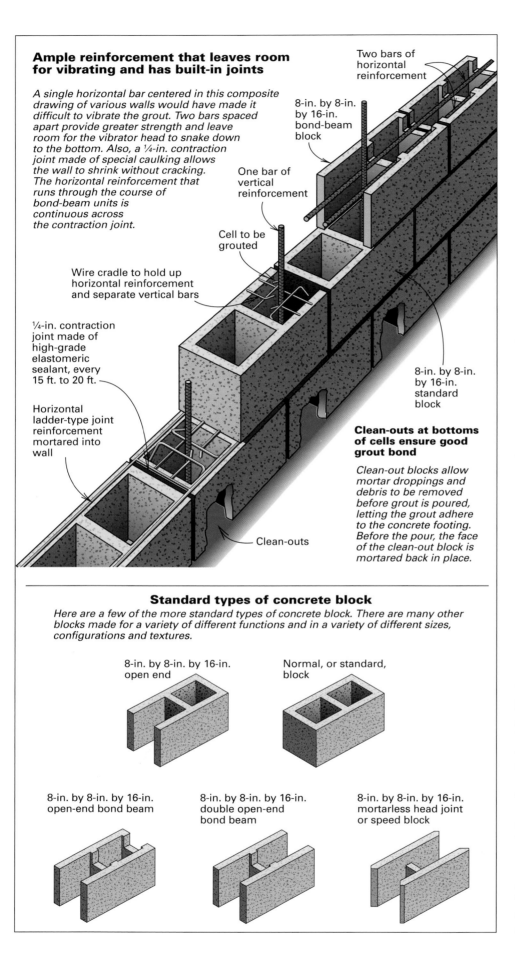

Ample reinforcement that leaves room for vibrating and has built-in joints

A single horizontal bar centered in this composite drawing of various walls would have made it difficult to vibrate the grout. Two bars spaced apart provide greater strength and leave room for the vibrator head to snake down to the bottom. Also, a ¼-in. contraction joint made of special caulking allows the wall to shrink without cracking. The horizontal reinforcement that runs through the course of bond-beam units is continuous across the contraction joint.

Two bars of horizontal reinforcement

8-in. by 8-in. by 16-in. bond-beam block

One bar of vertical reinforcement

Cell to be grouted

Wire cradle to hold up horizontal reinforcement and separate vertical bars

¼-in. contraction joint made of high-grade elastomeric sealant, every 15 ft. to 20 ft.

Horizontal ladder-type joint reinforcement mortared into wall

8-in. by 8-in. by 16-in. standard block

Clean-outs at bottoms of cells ensure good grout bond

Clean-out blocks allow mortar droppings and debris to be removed before grout is poured, letting the grout adhere to the concrete footing. Before the pour, the face of the clean-out block is mortared back in place.

Clean-outs

Standard types of concrete block

Here are a few of the more standard types of concrete block. There are many other blocks made for a variety of different functions and in a variety of different sizes, configurations and textures.

8-in. by 8-in. by 16-in. open end

Normal, or standard, block

8-in. by 8-in. by 16-in. open-end bond beam

8-in. by 8-in. by 16-in. double open-end bond beam

8-in. by 8-in. by 16-in. mortarless head joint or speed block

cores should be grouted after the wall is built, not as you go, which would mean cold joints.

Clean-outs help ensure that the wall bonds to the footing—Openings at the base of cells that receive grout allow you to remove construction debris before pouring (drawing left). When all of the blocks are laid, you can remove the mortar droppings, the nails, the tape measures, the cellular phones and the like that have fallen to the bottom of each cell. (Initially pouring an inch or so of sand at the base of the clean-out prevents fresh mortar from sticking to the footing. These chunks decrease the bond between the grout and the footing.) Then, seal the footings before pouring the grout, and let the mortar set a few days (to prevent blowouts) before the grout is poured. The UBC requires clean-outs in cells containing reinforcement if a grout pour will be more than 5 ft. high. For shorter lifts, you can suck debris out with a shop vacuum, although you may have trouble winding past all of those bars.

Ladder-type joint reinforcement is better than truss-type—Joint reinforcement can be used instead of bond-beam units that contain horizontal bars. This type of reinforcement consists of two horizontal, parallel wire rods connected by cross ties (drawing left). The reinforcement is placed between courses of block or brick, and the side rods get embedded in the mortar. The cross ties may go straight across or zigzag; these cross ties are called ladder and truss types, respectively.

The cross ties of ladder reinforcement should preferably align vertically with the webs of the blocks. Truss-type or zigzag-type wire reinforcement is even more difficult to align. We don't recommend using this reinforcement in concrete-block walls because the diagonal reinforcement can block the open cell and make it difficult to insert the vibrator.

Use pairs of horizontal rebar to make room for a vibrator—Rather than using single horizontal bars in bond beams, which lie in the middle of the wall cavity, use pairs of horizontal bars (either smaller bars or the same size at greater spacing). This process leaves more room for the vibrator head. Designers should consider this idea, but if yours hasn't, ask about it before proceeding. In straight runs of wall, open-end units allow grout to flow more easily through the block cavities. Using standard units forms air gaps between blocks at head joints.

An admixture that increases grout bond to block makes a stronger wall—Certain compounds react with grout ingredients to produce a gas. The gas expands, forcing the grout into the porous surface of the block cells. Several brands of admixture are available; most use powdered aluminum as the active ingredient (Grout-Aid by Sika Corp., 800-933-7452; or MB612 by Master Builders Inc., 800-628-9990).

Most masons add the admixture once the grout arrives at the site because its working time is limited to about an hour. Although the powder may be added straight into the truck, we require it first to be thoroughly mixed with water. If clumps of powder get pumped into the wall, they can generate enough gas pressure to pop the block apart.

Vibrate grout twice—For grout to flow into all of the crevices in a wall, it needs to be as fluid as possible. This wet grout is poured into dry masonry and vibrated immediately. The masonry soaks up excess water. As this soaking occurs, the grout loses volume. The grout may actually shrink away from one side of the grout space. The UBC requires reconsolidating the grout with a second vibrating. Vibrating a second time settles the grout fully into the cavities.

I have seen the level of a 4-ft. grout lift (the maximum height allowed) drop by 2 in. during revibration. It's important to wait at least 20 minutes before revibrating so that excess water has time to soak into the masonry. But do not wait so long that the expanding admixture reacts completely or that the grout begins to set up. In mild weather, revibrate within 45 minutes to an hour of placing the grout.

Vertical contraction joints reduce cracks in long walls—Masonry walls shrink in length and height as excess water in them dissipates. It may seem as if you're weakening the wall, but if you don't provide contraction joints to accommodate this shrinkage, the wall makes its own—in the form of cracks. To avoid shrinkage, build long walls in segments no more than 20 ft. long.

Build each segment as if it were an individual wall, but run horizontal reinforcement continuously. Separate the wall segments about ¼ in. Instead of a mortar joint between them, fill the gap with a high-grade elastomeric sealant, such as Sonolastic (Sonneborn; 800-433-9517) or Sikaflex (Sika Corp.; 800-933-7452). Joints can break up a wall's appearance, but they look better than cracks (photo p. 32).

Using some or all of these tips will make your work easier and stronger. For further information, the National Concrete Masonry Association (703-713-1900) publishes its TEK briefs on a wealth of masonry topics. ☐

Thor Matteson is a civil engineer in San Luis Obispo, California, and has taught part-time at California Polytechnic State University in San Luis Obispo. Photos by the author.

More about concrete reinforcement

A common gauge of steel's strength is its yield stress. When you apply stress below the yield stress and then release it, the steel returns to its original shape. But when you apply stress greater than the yield stress, the steel begins to deform permanently. The yield stress of reinforcing steel is indicated by its grade. Grade 40 means the yield stress is 40,000 psi. Most reinforcing bars are either grade 40 or grade 60.

U. S. steel mills produce reinforcing bars in 11 sizes, which are denoted by numbers (chart below). Those numbers represent the bar's nominal diameter in ⅛-in. increments. So a #4 bar is ½ in. dia. and a #18 bar is 2¼ in. dia.

Multiplying the bar's yield stress by its cross-sectional area gives its strength. For example, a #4 bar of grade-40 steel will withstand 8,000 lb. of force (0.2 sq. in. x 40,000 psi).

A bar's size and grade are stamped along its length, appearing in a column of symbols, letters and numbers. The first number to appear is the bar size. For some reason, the steel grade is more deeply hidden. Grade-40 bars have no special or additional marks. Bars of grade-60 steel have the number 60 stamped as the last symbol in the column, or they have an additional rib in their deformed pattern.

For the strength of a bar to be realized, it has to be held in a tight grip by the concrete. This bond strength comes from the bond between the concrete and the rebar, and it depends mostly on the concrete's strength and the holding ability of the rebar. The combination of

concrete strength and holding ability of rebar is known as development length. For a #4 grade-40 rebar, the correct development length is the distance the rebar must be embedded into the concrete so that when you pull on it with 8,000 lb. of force, the rebar stretches rather than pulls out of the concrete.

The American Concrete Institute's (ACI) model code gives the development length for small bars (#7 and smaller, typically) as 0.03 times the rebar's diameter in inches times its yield stress in psi divided by the square root of the concrete's compressive strength (in psi) at 28 days, the final result ending up in inches.

For more information on rebar, contact the Concrete Reinforcing Steel Institute, 933 N. Plum Grove Road, Schaumburg, Ill. 60173-4758; (708) 517-1200.—*T. M.*

Good information right on the bar. The *M* signifies the mill that made the steel; the *6* means it's #6 rebar; the *S* means it's made of new billet steel (the type normally found in rebar); the *6* and *0* mean its yield stress is 60,000 psi.

Standard reinforcing-bar sizes

To determine the strength of a given size of rebar, multiply its yield stress (40,000 psi and 60,000 psi for grades 40 and 60) times its cross-sectional area.

Bar size	Weight per ft. lb.	kg	Diameter in.	cm	Cross-sectional area in.	cm
#3	0.376	0.171	0.375	0.953	0.11	0.71
#4	0.668	0.303	0.500	1.270	0.20	1.29
#5	1.043	0.473	0.625	1.588	0.31	2.00
#6	1.502	0.681	0.750	1.905	0.44	2.84
#7	2.044	0.927	0.875	2.223	0.60	3.87
#8	2.670	1.211	1.000	2.540	0.79	5.10
#9	3.400	1.542	1.128	2.865	1.00	6.45
#10	4.303	1.952	1.270	3.226	1.27	8.19
#11	5.313	2.410	1.410	3.581	1.56	10.07
#14	7.650	3.470	1.693	4.300	2.25	14.52
#18	13.600	6.169	2.257	5.733	4.00	25.81

Operating a Power Trowel

Pressure on the handlebars controls the machine, and the angle of the blades controls the finish

by John M. Schnittker

I'll never forget the first time I used a power trowel. It was a big machine with a 48-in. dia. blade and a 12-hp engine. Luckily, plenty of experienced help was around to repair the damage done when the machine led me into an area of concrete that was still too wet to finish. What a mess. My short lesson taught me that you must know what you're doing before you use a power trowel. Here, I'll offer a few tips on operating the machine.

Opting for a power trowel—A power trowel, also called a troweling machine or rotary float, is a motor-driven concrete-finishing machine with revolving blades that smooth, flatten and compact a concrete slab (photo right). These tools are usually found on large concrete jobs. Most crews I've worked with used power trowels to finish slabs over 2,000 sq. ft.; for smaller pours, kneeboards and hand trowels were used. Still, the speed of a power trowel might be right for a smaller job, especially if you've got a small crew, or if you're pouring in hot weather or with a rainy forecast. And because a power trowel saves time on the finishing phase, it eliminates the temptation to add more water to the concrete during the pour, which makes for a better quality job. The relatively low rental cost of a power trowel in my area (between $40 and $60 per day) is easily offset by labor and time savings, tilting the balance toward a power trowel as an alternative to hand finishing.

A power trowel will not decrease the effort expended on placing, screeding and bull floating concrete. The machine saves time and effort only after the concrete is placed and leveled. A power trowel can float or finish 1,000 sq. ft. in about 20 minutes: about as fast as three or four good concrete finishers working on kneeboards.

Hand finishing still required—Basic concrete hand-finishing skills are still required, however, because a power trowel cannot finish flush against forms, get into corners or finish around recessed floor drains. So if you've never finished a concrete slab by hand, don't run out and rent a power trowel just yet. Practice hand finishing a few small pours—say between 250 sq. ft. and 500 sq. ft.—to ensure that you possess the necessary skills before you move up to power troweling. Besides, if the machine were to break down, your experience hand finishing concrete could save the day. Alternatively, consider rent-

Note the footprints. **This 36-in. power trowel has four rotating combination blades that float and finish a concrete slab. The operator walks backward and erases his footprints with the machine.**

ing a backup machine. Either way, make sure the machine starts at the rental yard before you bring it on site.

Learning the controls—The best power trowel for a beginner is also the smallest and lightest: a 5-hp, 36-in. machine fitted with combination float/trowel blades. This machine is more manageable than the powerful 42-in. and 48-in. power trowels, and it produces a smooth, hard finish perfectly acceptable in residential and light commercial applications. However, if a glasslike, burnished finish is desired, a larger, more powerful machine fitted with longer, narrower trowel blades will be needed.

Familiarize yourself with the controls before you begin finishing operations. The throttle, the blade-pitch control knob and the safety shutoff switch are on the handlebars. The throttle lever is on the right handlebar. The blade-pitch control knob is in the center of the handlebar. Turning the knob increases or decreases the pitch of the blades. Slight tension should always be main-

tained on this knob, even when the blades are flat, as they are during the floating operation.

Most power trowels have a safety shutoff switch that automatically turns off the machine when the handlebars are released. If the handlebars are released while the blades are turning, the blades will stop turning, and the rest of the machine—engine and handlebars—will start rotating. The safety switch shuts off the machine within one-quarter revolution; however, momentum and the disengaged clutch allow the handle to spin around a few times before everything comes to a stop.

Fire 'er up—Make sure the machine has been filled with the proper amount of fuel and oil before you place it on the slab. Spilling gasoline or oil on a concrete slab will cause discoloration and deterioration. Refueling, if required, should take place off the slab and between passes (each time you use the machine with a progressively higher blade pitch is called a "pass") after the machine has cooled down.

A power trowel is big and heavy, and it takes two people to tote one around. The machine is carried by the handlebars and by the housing around the blades, so you should never start the machine until it's been placed on the slab.

Many power trowels now have a centrifugal clutch, which allows the machine to be started and idled; blade engagement occurs as the throttle is opened. Also, most machines operate at blade speeds of 60 rpm to 150 rpm; the faster the speed, the smoother the finish. Sufficient throttle must be given to ensure full clutch engagement because a partially engaged clutch damages the machine. Older machines may be fitted with a mechanical clutch that is engaged by the operator. Like the safety shutoff switch found on new machines, a mechanical clutch disengages the blades when the handlebars are released.

Steered with handlebars—A few tips should keep you from repeating my first-time experience. At least half throttle should be applied when you are ready to start troweling. And low pitch angles on the blades are essential while getting a feel for the machine. You might consider practicing on a cured concrete slab that's been wetted down. A garage floor with a smooth, hard finish will do nicely, but a driveway with a coarse or a broomed finish will not: A rough surface might damage the blades.

Steering the power trowel is the interesting part (drawing right). With the machine running and the blades perfectly level, the machine stays put. Lift the handlebars up, and the machine moves to the left; push the handlebars down, and the machine moves to the right. Moving forward or backward is achieved with a slight twist of the handlebars. A twist to the left causes the machine to move forward; a twist to the right causes it to move backward.

If you find yourself wrestling the machine, you're overcontrolling it. Try to steer the machine with less effort. Even with low pitch angles on its blades, a power trowel responds to very light, almost fingertip pressure. After a few tries you should be comfortable controlling the machine.

The ease of control also depends on the slab itself because a bumpy slab causes the machine to jump and change speeds. Proper screeding and bull floating prior to finishing are always critical, regardless of whether you plan to finish the slab by hand or with a machine. A power trowel will not compensate for a poorly prepared slab.

Floating and troweling—It generally takes three passes with a power trowel to get a smooth, hard finish on a concrete slab. A fourth pass may be desired to attain a smoother, harder finish. If you're finishing a garage slab, remember that the floor gets wet and oily, so a glassy finish could invite accidents.

The first pass over the concrete is the floating pass and is accomplished with the blades set nearly flat, up to a couple of degrees. Floating can begin when the concrete has set up to the point where the machine and the operator leave only a slight depression on the surface of the slab—just the same as if you were going to finish the slab working from kneeboards with hand tools. The primary purpose of floating is to smooth the surface of the concrete. Small bumps are cut down, and small holes are filled with repeated passes over an uneven area. Floating should end once the concrete is smooth and flat.

Often, beginners overfloat the slab. Impressed by their first-time efforts, they decide to go over the concrete again while it is still in the wet, working stage. Overfloating pulls additional fines and excessive water to the surface and may result in hairline cracks or fractures after the concrete has completely set up. Try to allow the concrete to harden following a single floating pass.

The second pass is really the start of the troweling operation. You can tell when the slab is ready for the second pass by working the edges with a hand trowel. Once the concrete accepts a smoother finish, power troweling can begin.

With a 36-in. machine, there's no need to change the blades; the combination blades are designed for floating and troweling. Bigger machines have trowel blades; float blades clip onto the trowel blades and must be removed before the troweling operation begins.

For the second pass, turn the blade-pitch control knob so that the blades are set at about 5°. Pitching the blades upward puts more pressure on the slab. As you increase the pitch of the blades, the finish becomes smoother, and the machine is more responsive to operator input.

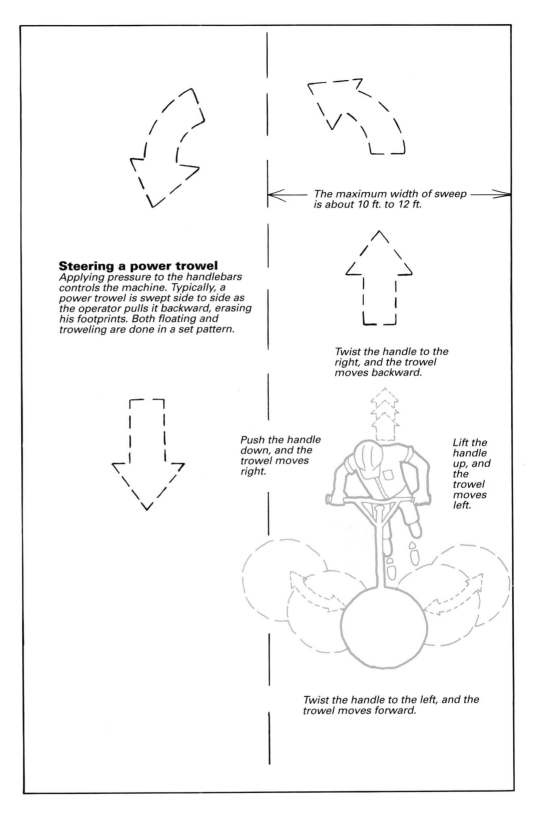

Steering a power trowel
Applying pressure to the handlebars controls the machine. Typically, a power trowel is swept side to side as the operator pulls it backward, erasing his footprints. Both floating and troweling are done in a set pattern.

The maximum width of sweep is about 10 ft. to 12 ft.

Twist the handle to the right, and the trowel moves backward.

Push the handle down, and the trowel moves right.

Lift the handle up, and the trowel moves left.

Twist the handle to the left, and the trowel moves forward.

Subsequent passes over the concrete occur with increasing pitch angles (10° to 20°) and depend on how smooth and hard a finish you desire.

Both floating and troweling should be done with a set pattern. Generally, the power trowel is swept right to left and left to right while moving backward. A single right-to-left sweep covers 10 ft. or 12 ft. and takes approximately 15 seconds. The operator walks backward, troweling out his footprints and checking that the area behind him is sufficiently hard for floating or troweling. How quickly you move backward depends on how much you overlap previously troweled areas. I like to overlap approximately 50% of the sweep (18 in. on a 36-in. machine), which results in the power trowel going over the slab twice during a given pass. Forward movement of the power trowel occurs when working the machine into a corner or when making repeated back-and-forth passes to smooth out an area during floating. □

John M. Schnittker is a resident of Fort Collins, Colo. Photo by Rich Ziegner.

Drawing: Mary Twellmann

Rubble-Trench Foundations

A simple, effective foundation system for residential structures

by Elias Velonis

Although it was first used extensively by Frank Lloyd Wright early in the 20th century, the rubble or gravel-trench foundation has largely been ignored by builders since Wright's time—perhaps because it represents a different way of thinking about what it takes to support a house. The conventional poured-concrete or block perimeter wall attempts to solve a building's load-bearing requirements in monolithic fashion by creating a solid, supposedly immovable and leakproof barrier extending from a footing poured below frost line to 8 in. or more above grade. But since freezing water expands 9% by volume with a force of 150 tons per sq. in., monolithic foundations are unlikely to survive in frost country unless they include a footing-level perimeter drain backfilled with washed stone, which carries away water that might collect and freeze under or against the foundation wall.

The two functions of load-bearing and drainage are solved separately with a solid foundation, but the rubble-trench system unites these two functions in a single solution: the house is built on top of a drainage trench of compacted stone that is capped with a poured-concrete grade beam. The grade beam is above the frost line, but the rubble trench extends below it, and the building's weight is carried to the earth by the stones that fill the trench (see drawing, facing page, center). The small airspaces around each stone allow groundwater to find its way easily to the perforated drainage pipe at the bottom of the trench. Atop the grade beam, a short stemwall of concrete block, poured concrete or pressure-treated wood is built to support the floor framing. Or you can pour a slab. More about this later.

While this foundation system has been time-tested in many of Wright's houses, acceptance by building officials and the codes they follow is still not assured. In *The Natural House* (Horizon Press, New York, 1954), Wright speaks of what he calls the dry wall footing. "All those footings at Taliesin have been perfectly static. Ever since I discovered the dry wall footing—about 1902—I have been building houses that way.... Occasionally there has been trouble getting the system authorized by building commissions."

The disapproval of a building inspector usually arises from a lack of familiarity with the technique, since the Uniform Building Code states clearly that any system is acceptable as long as it can "support safely the loads imposed." When I first approached our local building inspector with plans for a rubble-trench foundation, he studied them quietly for a moment, ahemmed in good New England fashion, and said, "Yep, that looks as if it oughta work." And so it will, except in what Wright calls "treacherous soils," which I would judge to be any soils with a bearing capacity of less than 1 ton per sq. ft.

Determining the bearing capacity of a soil without engineering analysis is a matter of common sense and experience. If the earth in the trench is dry, seems to be well drained, feels solid when you jump on it, and is a mixture of gravel, rock, sand or clayey sands, it will very likely carry all the weight your house can bear on it. If, on the other hand, your heels sink several inches into soft clay, loose sand or fine silt when you jump into the trench, you'd better consult a soils engineer.

Construction—Assuming you've got stable soil, bulldoze the area of the house level, clearing all topsoil away and saving it for fin-

====

'All those footings at Taliesin have been perfectly static. Ever since I discovered the dry wall footing—about 1902—I have been building houses that way.' —Frank Lloyd Wright

====

ish grading. If you have a sloping site, you will have to cut a level shelf in the hill, graded away from the house on all sides. This will ensure a good path for surface runoff. Lay out your foundation in the conventional manner (see "Site Layout," pp. 17-19), but make sure the batter boards are set up far enough outside the lines of the building that the backhoe will have room to maneuver. Sprinkle a line of lime 4 in. inside the strings that define the building's outer edge. This white line represents the center of the masonry wall that will rise up from the on-grade footing, or grade beam, and it provides the backhoe operator with a centerline to follow with his bucket. Ask the excavator if he has a narrow bucket for the backhoe—16 in. to 20 in. is perfect for

most soils. A wider trench gives you more bearing in softer soils, but it also takes more stone to fill it.

Have the backhoe operator cut the trench with straight sides, as deep as the frost line at the high point and sloping down to one or more outlet trenches along the perimeter (see drawing, facing page, bottom). These should run away from the building and out to daylight at a slope of at least 1 in. in 8 ft. If you have a level site, I recommend running trench drains to a drywell, if your water table isn't too high. A drywell is a hole filled with a combination of small (1½-in.) stone and coarser rubble. You can base the depth and diameter of your drywell on the drainage qualities of your soil and the surface runoff you expect. Compute this from average-rainfall data and figures from the site's percolation test.

Clean up all your trenches by hand, making sure that their bottoms are flat and that they slope toward the drain line. Disturbed soil at the bottom of the trench may settle unevenly, so tamp the bottom firm with a pneumatic tamper or the heels of many boots.

Next, pour in a few inches of washed stone, and lay 4-in. dia. perforated PVC drainage pipe on top of it in the foundation and outlet trench. Make sure that the pipe follows the slope without dips that could restrict the flow of water. A ½-in. block taped to the end of a 4-ft. level makes the job of sloping the rigid pipe quite a bit easier. When the bubble reads level, you've got a 1-in. in 8-ft. slope.

I place the perforated pipe with its holes down (that is, at 4 o'clock and 8 o'clock), as I would in laying out a leach field, because as the trench fills with water, I think this orientation gets rid of it quicker. On the other hand, a case could be made for putting the holes up. It would take longer for them to silt up, but this shouldn't be a problem in good soils.

Now begin filling the trench with washed stone, taking care not to disturb the pipe as you cover it. I use 1½-in. stone because it's easy to find and easy to shovel, but larger washed stone is okay, too, as is the occasional clean fieldstone. (This is where the technique gets the name rubble trench.) Tamp the stone every vertical foot or so to make sure it is compact. To this end, I have even driven a loaded dump truck along the filled trench to make sure it was well settled, although this seemed to have little effect.

The outlet trench need not be filled with

Washed stone is dumped straight from the truck into the foundation and outlet trenches.

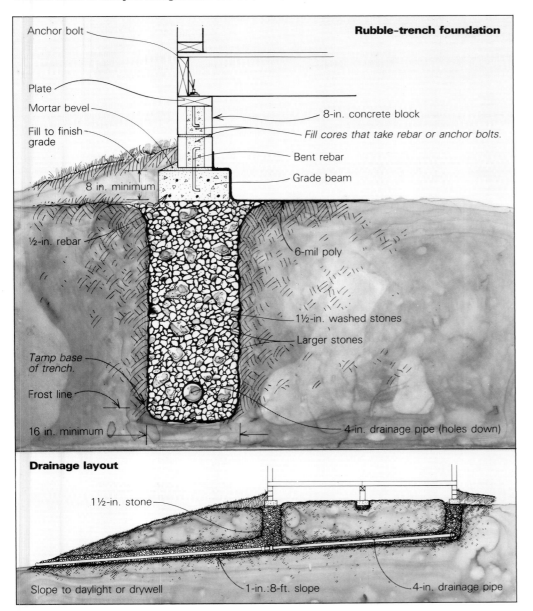

Rubble-trench foundation

Anchor bolt

Plate

Mortar bevel

Fill to finish grade

8 in. minimum

½-in. rebar

Tamp base of trench.

Frost line

16 in. minimum

8-in. concrete block

Fill cores that take rebar or anchor bolts.

Bent rebar

Grade beam

6-mil poly

1½-in. washed stones

Larger stones

4-in. drainage pipe (holes down)

Drainage layout

1½-in. stone

Slope to daylight or drywell

1-in.:8-ft. slope

4-in. drainage pipe

Illustrations: Jackie Rogers

stone except for a foot in all directions around the pipe. Cover this stone with hay, burlap or tar paper as a filter, and backfill it with the original soil. If the pipe is running to daylight, be sure to leave its end exposed on a bed of stone. You want it to drain freely, so don't cover it with soil. Cap the end of the pipe with wire mesh to keep out rodents.

The grade beam—After the drains are installed and the trenches filled with stone, you're ready to build the forms for the grade beam. For one-story wood-frame structures, a 16-in. wide by 8-in. deep grade beam with three runs of ½-in. rebar is more than adequate. For a two-story structure, increase the depth of the beam to 10 in. or 12 in., and add two more lengths of rebar in its upper third.

Restring the lines from the batter boards and place the form boards on edge beneath them. I use 2x8s or 2x10s and brace them with stakes every 3 ft. Level the top edges of the form boards all around, nailing in 1x2 spreaders every 4 ft. to 6 ft. across the top to hold the forms in place. To reinforce the corners, use metal strapping or plumber's tape (perforated steel strapping), nailing it around outside corners. Place three runs of ½-in. rebar spaced evenly along the bottom of the beam, wiring them securely at the joints, and stagger these joints around the perimeter. Wire short pieces of rebar across these runs every 6 ft. Bend the lengths of rebar around all corners rather than splicing them there. Lift the rebar about 2 in. off the bottom of the beam with small stones. For a two-story building, prepare two more perimeter runs of rebar, and put them alongside the forms, ready to be dropped in the top third of the beam during the pour.

As the concrete is poured into the form, vi-

Forming and pouring the grade beam. At right, forms for the grade beam are being set up on top of the stone-filled trenches. The 2x10 form boards are held together by steel plumber's strapping and by 1x2 wood stretchers nailed across their tops. Below right, rebar has been placed between the forms, and tied off to spreaders. The transit mixer is discharging its load of concrete, which is being spread and leveled by the crew of Heartwood students.

brate it well with a short piece of 2x4 to get rid of air pockets. Screed along the tops of the forms to get a level surface. If you intend to build a masonry wall, rough up the top of the grade beam with a broom before the concrete cures to ensure a good bond between it and the mortar. In areas where high winds are a problem, set 1-ft. lengths of bent rebar vertically in the top of the grade beam. They will anchor the stemwall. For a pressure-treated stemwall, place anchor bolts for the sill plate every 6 ft., and 1 in. from the end of each plate member.

If your design calls for a slab floor, you can pour the grade beam and slab at the same time, using a turned-down or "Alaskan" slab. For this you need only build outside forms, as shown in the second drawing below. For

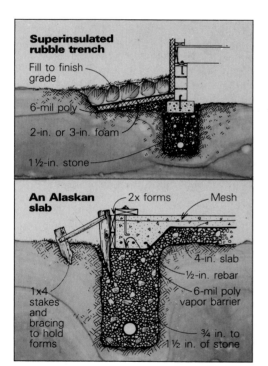

Superinsulated rubble trench
Fill to finish grade
6-mil poly
2-in. or 3-in. foam
1½-in. stone

An Alaskan slab
2x forms
Mesh
4-in. slab
½-in. rebar
6-mil poly vapor barrier
1x4 stakes and bracing to hold forms
¾ in. to 1½ in. of stone

The finished grade beam, with its forms stripped.

houses with wooden floors, however, the height of the stemwall will determine the height of the crawl space, which should be at least 26 in. Since this may make the level of the finished floor higher than you want, raise the grade around the perimeter by 1 ft. or so, which will help slope it away from the house for surface runoff. Leave adequate ventilation ports on the stemwall, and when the building is roofed, cover the earth in the crawl space with 6-mil poly. Place anchor bolts for the plate in the usual manner, and parge the outside of the wall that will be above grade for a clean appearance. You might also add a bevel of mortar between the stemwall and the grade beam, to shed any water that might want to find its way between them.

Insulation alternatives—If you've designed an energy-efficient building and want to extend the insulation down below grade on the exterior, modify the foundation as shown in the first drawing at left. In an area where the frost line is 4 ft. deep, dig the trench only 2 ft. to 2½ ft. below initial grade, as you will be raising the finished grade around the building by 1½ ft. to 2 ft. Fill the trench with stone to 8 in. below initial grade, and then pour the grade beam to the surface. Lay up four courses of block (surface-bonded block walls work very well in this application) and build the floor.

When you're ready to insulate the exterior of the building, glue 2-in. or 3-in. thick panels of rigid foam insulation to the exterior of the stemwall down to the grade beam, then lay more foam panels on a sloping bed (at least 3 in. in 1 ft.) of stone, 2 ft. to 4 ft. around the entire perimeter. This insulation apron preserves a large bubble of relatively warm earth beneath the house, tempering the crawl space with its warmth. Cover this apron with 6-mil poly to protect the foam from water, and backfill up to finish grade. The exposed foam above grade and below the siding should be covered with asbestos board or the equivalent, or parged (see pp. 59-61) with a surface-bonding compound troweled on ⅛ in. thick over a wet coat of Styrofoam adhesive. The adhesive helps eliminate expansion cracks in the surface bonding between panels, but if cracks do appear, they can be caulked.

The only reservation I've heard from other builders about the rubble-trench foundation is that it might settle unevenly. In non-uniform soils this might be a problem, although the reinforcing in the grade beam is ample to span a good deal of uneven settling. We've built four rubble-trench foundations over the last five years, and none has shown the slightest sign of settling, cracking or frost damage. The great advantage of the system, of course, besides speed and relatively low cost, is that instead of building a massive wall underground, you just pour stones into a trench and are free to carry on building above ground. □

Elias Velonis is the founder and co-director of Heartwood Owner-Builder School in Washington, Mass. Photos by the Heartwood staff.

Stepped Foundations
Using modular forms to build a foundation on a hillside

by Michael Spexarth

To the builder who's about to put in a foundation, building sites can pose a variety of problems. Assuming that soil conditions are stable, the flat lot with easy access is at the no-problem end of the scale, while the 10-in-10 (45°) "view lot" occupies the nightmare position. The standard spread-footing/stemwall foundation is the most common footing for the easy site. Here in the San Francisco Bay Area, the pier-and-grade-beam foundation is generally used on the steep ones (see pp. 62-65).

A lot of sites fall between these two extremes, and confronted with one of these, I like to build a modified spread-footing foundation. This is called a stepped foundation, and building one is standard procedure on a site that has stable, well-drained soil with a slope between 2 in 10 and about 5 in 10. I would seek an engineer's advice on any slope greater than 5 in 10, for soil conditions and steel requirements.

Stepped foundations take on the appearance of staircases as they stretch out across a lot, and their level changes make them look like the product of very complicated formwork. Well, they are complicated, but I avoid needless complexity by using modular form panels that can be used over and over again.

The panels—When I built the foundation in the photo below, I used 2-ft. by 8-ft. panels made from 2x4s and ½-in. plywood. The most durable forms are made with exterior-grade AC plywood, with the A side in contact with the concrete. The 2x4s are used on edge to reinforce the plywood panels around the perimeters, and on 16-in. centers in the field. Panels that are 8 ft. long are perfect for stepped foundations on hillsides that have a slope of about 2 in 10, while 4-ft. by 2-ft. panels are appropriate for a slope approaching 5 in 10 (top drawing, below). If your site is in the 2-in-10 range, the 8-ft. panels are especially handy if the outside perimeter of your foundation or retaining wall is divisible by 8, such as 24 ft. or 48 ft.

The inside panels are the same length as the outside ones, except at the corners. There an allowance has to be made for the thickness of the stemwall you are forming—8 in. for a two-story building, 6 in. for a one-story building, according to the Uniform Building Code. The foundation shown here supports a one-story house, so the inside-corner form panels had to be 7 ft. 6 in. in the north-south direction, and 7 ft. 2 in. in the east-west direction (bottom drawing).

Although it would be nice if foundation plans and hillside contours all came in 8-ft. increments, they don't, and a builder has to improvise when the form panels don't quite match the site. If you have an odd-sized foundation, a foundation with a lot of corners or a site with grade changes, you will have to cut or add small sections to panels, usually at the corners. Fortunately it's easy to add a smaller panel to act as a spacer or an extension when needed.

Excavation—When you plan the excavation, you have to decide how long the excavation cuts, or shelves, should be for the footings. For the 2-in-10 slope under this foundation, 8-ft. panels over 8-ft. shelves worked out fine. If a variation in the slope makes it necessary to cut shelves of different lengths, you can gang panels together or let a panel overlap the one below it for an adjustment.

Once we had our batter boards positioned, we stretched string lines about a foot off the ground to mark the centerline of the foundation. Then we shook handfuls of white lime over the string lines to mark their position on the ground. Most backhoe operators tend to wander right or left, depending on whether they are right or left-handed, and on sloped sites they have to con-

Plywood panels *reinforced with 2x4s are used by the author to form stepped foundations on sloping sites. Opposing pairs of panels hold concrete for the stemwalls—the footings below them are formed in excavated trenches. At the downhill end of each step, a plywood dam tacked to the panels keeps the concrete on the right level.*

5-in-10 slope

Average footing depth

4 ft.

2 ft.

Wire corners together.

8 ft.

6 in.

7 ft. 6 in.

Interior panels

7 ft. 2 in.

8 ft.

Exterior panels

Plan of corner

sider the potential for rolling their equipment. The limed line is their reference point, and allows them time to consider their position on the hill. They get safer use of their equipment, and we get a straight excavation, which cuts down on the hand-digging.

On this job, we had the operator use a 24-in. bucket, and every 8 ft. we had him drop down 2 ft., leaving 24-in. wide level steps 8 ft. long in the ground. The shallowest point in each trench should be equal to the depth of the footing required by the building and the climate. Don't expect that the shelves will be exact in length or that they will be spot-on level. Such adjustments are made as the form panels are placed.

After the steps had been dug by the backhoe, we ran string lines from our batter boards to mark the outside edges of the stemwall. Then we set the outside panels to the line marked by the string. Where the slope or the excavation made large gaps at the lower edge of the panels, we nailed a plywood skirt to hold in the concrete, as shown in the drawings below. We also closed up the open ends of the forms with plywood dams.

The panels are held in place by nails driven through steel stakes into the 2x4s along the panel tops and bottoms (photo facing page). Three stakes per panel and two nails per stake are enough to secure them. I like to use steel stakes for most foundation work because they are easier to drive into hard soil, loose rock, sandstone or shale. They are also easier to remove than wooden stakes because they exert less friction against the soil.

After checking to make sure that the outside forms were square and plumb, we added 2x4 kickers (or braces) to stiffen them. Most panels had three kickers, and we made sure that the highest points of the forms, near the steps, were well-braced. Once the outside panels were secure, we set the inside forms. We started in the middle of a run, and worked toward the corners. The inside panels matched the length and height of the outside ones, and we nailed plywood ties across their tops every 3 ft. or so to keep the distance between the panels at a steady 6 in. To set the distance between the panels along their middles, we cut 6-in. lengths of 2x4 to act as spacers. With these in place, we ran tie-wire loops between the two panels and twisted them tight. Then we removed the spacers. The loops keep the forms from spreading too far as they are filled.

Steelwork—Our plan called for two pieces of #4 rebar at the top and at the bottom of the foundation. In a stepped foundation, the steel is also stepped. Horizontal bars become vertical at the step, then horizontal again. Given the number of bars embedded in a 6-in. wall, things can get crowded at the step. I space the bars so they will be evenly distributed at this point, 2 in. from the face of the concrete and 2 in. from each other. If the panels are overlapping by 12 in. to 18 in., I add one or two vertical bars. The step is the weakest point in this type of foundation, so it's important to have plenty of steel there.

Since a stepped foundation means stepped

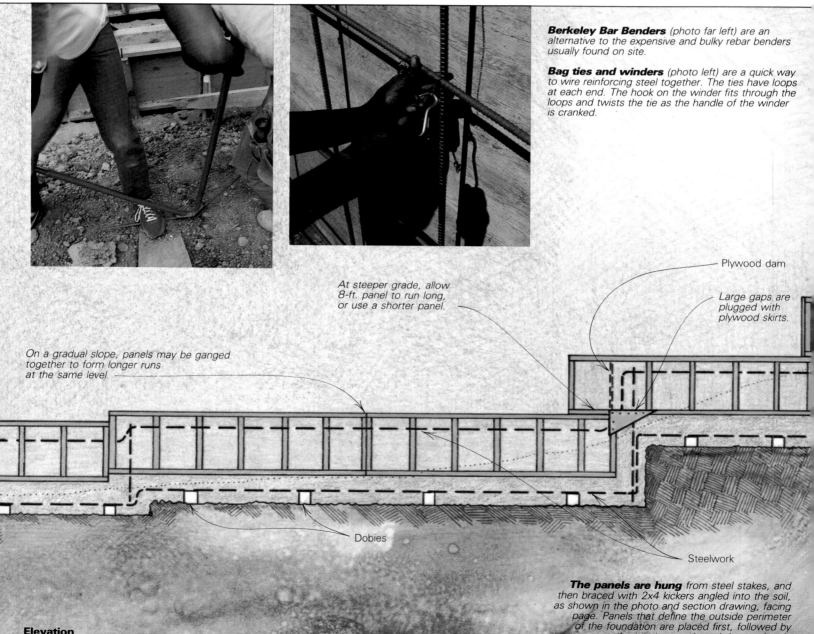

Berkeley Bar Benders (photo far left) are an alternative to the expensive and bulky rebar benders usually found on site.

Bag ties and winders (photo left) are a quick way to wire reinforcing steel together. The ties have loops at each end. The hook on the winder fits through the loops and twists the tie as the handle of the winder is cranked.

Plywood dam

Large gaps are plugged with plywood skirts.

At steeper grade, allow 8-ft. panel to run long, or use a shorter panel.

On a gradual slope, panels may be ganged together to form longer runs at the same level.

Dobies

Steelwork

Elevation

The panels are hung from steel stakes, and then braced with 2x4 kickers angled into the soil, as shown in the photo and section drawing, facing page. Panels that define the outside perimeter of the foundation are placed first, followed by the interior panels and the steelwork.

steel, there is a lot of rebar bending in a job like this. We used Berkeley Bar Benders to put the steps in the steel (photo facing page, far left). They are two 3-ft. lengths of pipe that have a 3-in. length of pipe welded onto their ends at a 60° angle. These 3-in. lengths are slotted to accept a piece of #4 rebar. One worker uses a bender to hold the rebar steady while another levers the bar to the desired angle. If you don't have to fold rebar week in and week out, the benders, manufactured by Berkeley Bar Bender (1215 Shattuck Ave., Berkeley, Calif. 94709), are an inexpensive (about $40) alternative to the heavy-duty cutter-benders that are used by concrete specialists.

We stepped the lower pair of bars to match the stepped panels, and tied them to dobies, which are precast 3-in. cubes of concrete. The dobies are attached with tie-wire to the rebar, and they keep the rebar 3 in. above the ground, as required by code.

Since we had plenty of rebar that needed tie-wire on this job, we used bag ties and a winder to speed things along. Bag ties are short cutoffs of foundation tie-wire (6 in., 8 in. or 10 in. long) with a loop on both ends. They come strung together in rolls of 1,000. To use one, you wrap it around the steel with one hand, and insert a winder through the two loops. A winder is a steel hook that's free to spin in a wooden handle. You crank on the handle, and the hook winds the wire and the steel together (photo facing page, center). No more tangled rolls of tie-wire or awkward angles for pliers.

We pump-poured this foundation because it was well off the street, but the forms would take any mix. I start a pour with one pass around the perimeter to fill up the footings. Then I fill the forms halfway and check them for any distortion, separation or split wood. If I have a problem at this point, I know I'm going to have a disaster later, so I stop the pour while I sort things out. This accomplishes two things: I can fix the questionable formwork, and the concrete has time to set up and carry its own weight, which takes some of the burden off the forms. If

I have several workers on the job, they continue the pour at other areas of the foundation.

Once the forms are topped off, we rough-screed the stemwalls using the top of the forms as a guide. Then I set the bolts. As the mix sets up, we remove the plywood ties across the forms. The stemwall top can then be finished as desired. This is a good time to pull out the steel stakes—you've got about 24 hours before they become a permanent part of the foundation.

If you carefully scrape the panels and coat them with form-release after every job, they will eventually pay for themselves. Form-release is an oil-base coating manufactured by companies that sell concrete accessories, such as the Burke Company (2655 Campus Drive, San Mateo, Calif. 94403). Most small contractors just use old motor oil. Around here, many builders rent out their form panels to help each other out, and to defray costs. □

Michael Spexarth is a contractor and part-time building inspector in El Cerrito, Calif.

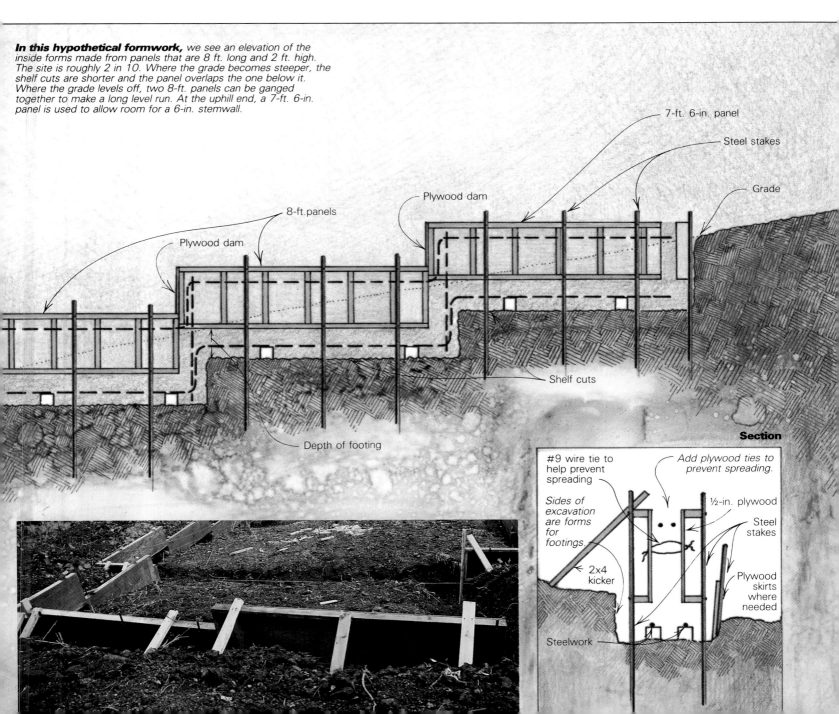

In this hypothetical formwork, *we see an elevation of the inside forms made from panels that are 8 ft. long and 2 ft. high. The site is roughly 2 in 10. Where the grade becomes steeper, the shelf cuts are shorter and the panel overlaps the one below it. Where the grade levels off, two 8-ft. panels can be ganged together to make a long level run. At the uphill end, a 7-ft. 6-in. panel is used to allow room for a 6-in. stemwall.*

7-ft. 6-in. panel

Steel stakes

Grade

Plywood dam

8-ft. panels

Plywood dam

Shelf cuts

Depth of footing

Section

#9 wire tie to help prevent spreading

Add plywood ties to prevent spreading.

Sides of excavation are forms for footings.

½-in. plywood

Steel stakes

2x4 kicker

Plywood skirts where needed

Steelwork

Building a Block Foundation

How to pour the footing and lay the concrete block, and what to do about waterproofing, drainage and insulation

by Dick Kreh

Foundation walls are often the most neglected part of a structure. But they are actually the most structurally important element of a house. They support the weight of the building by distributing its entire load over a large area. Apart from structural requirements, foundations have to be waterproofed, insulated and properly drained.

Although the depth of a foundation wall may vary according to the specific needs of the site or building, the footings must always be below the frost line. If they're not, the foundation will heave in cold weather as the frozen earth swells, and then settle in warm weather when the ground softens. This shift-

ing can crack foundations, rack framing, and make for wavy floors and sagging roofs.

Concrete blocks are composed of portland cement, a fine aggregate and water. They have been a popular choice for foundations because they're not too expensive, they go up in a straightforward way, and they're available everywhere. Block foundations provide adequate compressive strength and resistance to fire and moisture. They don't require formwork, and they're not expensive to maintain.

All standard blocks are 8 in. high and 16 in. long—including the usual ⅜-in. thick mortar head and bed joints. But they come in different widths. The size given for a block always

refers to its width. The size you need depends on the vertical loads and lateral stresses that the wall will have to withstand, but as a rule, most concrete-block foundations are built of 10-in. or 12-in. block.

Footings—After the foundation area has been laid out and excavated, the concrete footings are poured. Footings should be about twice as wide as the block wall they will support. A 12-in. concrete-block foundation wall, for example, should have a 24-in. wide footing. The average depth for the footing, unless there is a special problem, is 8 in.

Concrete footings for homes or small structures need a compressive strength of 2,500 pounds per square inch (psi). You can order a footing mix either by specifying a five-bag mix, which means that there are five bags of portland cement to each cubic yard, or by asking for a prescription mix—one that is ordered by giving a psi rating. Some architects and local building codes require you to state the prescription mix when you order. Either way, footing concrete is a little less expensive than regular finishing concrete, which usually contains at least six bags of portland cement to the cubic yard. The six-bag mix is richer and easier to trowel, but isn't needed for most footings. For more on mixing and ordering concrete, see pp. 8-13 and 108-109.

There are two types of footings—trench footings and formed footings. If the area where the walls are to be built is relatively free of rock, the simplest solution is to dig a trench, and use it as a form. Keep the top of the concrete footings level by driving short lengths of rebar to the proper elevation. Don't use wooden stakes because later they'll rot and leave voids in your footing. You'll need a transit level or water level to get the rods at the right height. After you install the level rods even with the top of the proposed footing, pour concrete in the trench, and trowel it flush with the tops of the rods. Some building codes require that these stakes be removed before the concrete sets up.

If the ground is rocky, you may have to set up wooden forms and brace them for the pour. I've saved some money in this situation by ordering the floor joists for the first floor and using them to build the forms. This won't damage the joists and will save you a lot of money. When the concrete has set, I remove the boards and clean them off with a wire brush and water. The sooner you remove the forms, the easier it will be to clean them.

After the footings have cured for at least 24 hours, drive nails at the corners of the foundation. To find the corner points, use a transit level or drop a plumb line from the layout lines that are strung to your batter boards at the top of the foundation (for more on laying out foundations, see "Site Layout," pp. 17-19). Next, snap a chalkline between the corner nails on the footings to mark the wall lines. Stack the blocks around the inside of the foundation. Leave at least 2 ft. of working space between the footing and your stacks of block. Also, allow room for a traffic lane so the workers can get back and forth with mortar and scaffolding.

Mortar mix for block—For the average block foundation, use masonry cement, which is sold in 70-lb. bags. You have to supply sand and water. Masonry cement is made by many companies. Brand name doesn't matter much,

but you will need to choose between mixes of different strength. The average strength, for general masonry work, is universally classified as Type N. Unless you ask for a special type, you'll always get Type N. I get Type N masonry cement unless there is a severe moisture condition or stress, in which case I would use Type M, which is much stronger. The correct proportions of sand and water are important to get full-strength mortar. Like concrete, mortar reaches testing strength in 28 days, under normal weather conditions.

To mix the mortar, use one part masonry cement to three parts sand, with enough water to blend the ingredients into a workable mixture. Mortar for concrete block should be a little stiffer than for brickwork, because of the greater weight of the blocks. You will have to experiment a little to get it right. The mortar must be able to support the weight of the block without sinking.

The mixing water should be reasonably clean and free from mud, silt or organic matter. Drinking water makes good mortar. Order washed building sand from your supplier. It's sold by the ton.

The following will help you estimate the amount of mortar you'll need: One bag of masonry cement when mixed with sand and water will lay about 28 concrete blocks. Eight bags of masonry cement, on the average, will require one ton of building sand. Remember that if you have the sand dumped on the ground, some will be lost since you can't pick it all up with the shovel. For each three tons, allow about a half-ton for waste.

Laying out the first course—Assuming the footing is level, begin by troweling down a bed of mortar and laying one block on the corner. Tap it down until it is the correct height (8 in.), level and plumb.

A block wall built of either 10-in. or 12-in. block requires a special L-shaped corner block, which will bond half over the one beneath. The point is to avoid a continuous vertical mortar joint at the corner. Now lay the adjoining block. It will fit against the L-shaped corner block, forming the correct half-bond, as shown in the drawing and photo at right.

When the second L-shaped corner block is laid over the one beneath in the opposite direction, the bond of the wall is established. On each succeeding course the L corner block will be reversed.

Once the first corner is laid out, measure the first course out to the opposite corner. It's best for the entire course to be laid in whole blocks. You can do this simply by using a steel tape, marking off increments of 48 in., which is three blocks including their mortar head joints. Or you can slide a 4-ft. level along

the footing and mark off 48-in. lengths. In some cases, of course, dimensions will require your using a partial block in each course, but it's best to avoid this wherever possible. If a piece of block must be used, lay it in the center of the wall or where a window or partition will be, so it is not as noticeable. After the bond is marked on the footing, a block is laid on the opposite corner and also lined up. Then you attach a mason's line to the outside corner and run it to the opposite corner point. This is called "ranging" the wall.

Sometimes there are steps in the footings because of a changing grade line. The lowest areas should always be built up first to a point

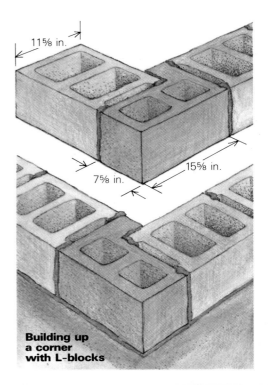

11⅝ in.

15⅝ in.

7⅝ in.

Building up a corner with L-blocks

Laying the first course begins at the corners. Once the four corners are laid and aligned, the entire bottom course is laid directly atop the footing. Most foundation walls are built from block 10 in. or 12 in. wide, and special L-shaped corner blocks, like the one shown here, have to be ordered. Only 8-in. wide blocks can be laid up without L-shaped corner blocks.

Illustrations: Christopher Clapp

where a level, continuous course of block runs through from one corner to the other. Steps in footings should be in increments of 8 in. so that courses of block work out evenly.

After one course of block is laid completely around the foundation to establish the bond and wall lines, it's time to build up the four corners. But before you begin laying block, you should make a story pole, sometimes called a course rod. Do this by selecting a fairly straight wooden pole and marking it off every 8 in. from the bottom to the height of the top of the foundation wall.

Any special elevations or features, such as window heads, door heads, sills and beam pockets, should be marked on the pole to coincide with the 8-in. increments wherever possible. After checking all your pencil marks, make them permanent by kerfing the pole lightly with a saw. Then cut the pole off even with the top of the foundation wall and number the courses of block from the bottom to the top so you don't find yourself using it upside down.

Now you can start laying up the corners so that you end up with only one block at the level of the top course. Successive courses are racked back half a block shorter than the previous ones (photo top left), so trowel on only enough mortar to bed the blocks in a given course. If the local code or your specifications call for using wire reinforcement in the joints, leave at least 6 in. of wire extending over the block. At the corners, cut one strand of the wire, and bend the other at 90°, rather than butting two sections together and having a break in the reinforcement.

Check the height of the blockwork periodically with the story pole. The courses of block should line up even with the kerfs. Once the corners are laid up, you can begin to fill in the wall between. Keep the courses level by laying them to a line stretched between the corners. Keep the corners plumb by checking every course with a spirit level.

Using manufactured corner poles—So far, I've described laying a foundation using the traditional method of leveling and plumbing. But in recent years, manufactured metal corner-pole guides have become popular with builders. They guide the laying up of each course and require less skill than the old way. They work like this. The corner poles are set on the wall once the first course of block is laid out. They are plumbed, then braced in position. Each pole has course heights engraved on it. Line blocks are attached to the poles on opposite corners at the desired course height, and the wall is laid to the line. There is no doubt that the use of manufac-

After the first course is laid, the corners are built up to the topmost course. Above left, a mason checks course heights against a story pole, which is graduated in 8-in. increments.

Reinforced concrete lintels, left, are used to tie the main foundation to walls that are laid at a higher level, such as porch foundations or garage walls.

tured corner-pole guides has increased the mason's productivity without adversely affecting the quality of the work.

If you have to tie a porch or garage wall into a main foundation at a higher elevation, lay a concrete-block lintel in mortar from the corner of the wall being built to the footing at the higher elevation (photo bottom left). Then lay blocks on the lintel to form the wall. This saves time and materials in an area that doesn't require a full-basement foundation.

Stepping the wall at grade line—As you build up to the natural grade line of the earth, you can set the front of the wall back about 4 in. to form a shelf for a brick veneer, if the plans call for it. This is done by switching to narrower block—from 12-in. block to 8-in. block, or from 10-in. block to 6-in. block. The inside of the wall stays in the same plane.

Making the last course solid—On some jobs, specifications require that the last course of block be solid to help distribute the weight of the structure above and to close off the holes. You need only grout the voids in the top course of block. Broken bits of block wedged into the voids in the course below will keep the concrete from falling through. The sill plates will rest on this top course, and the floor joists on top of the plate.

The sill plate has to be bolted down to the top of the foundation wall. So you have to grout anchor bolts into the top of the wall every 4 ft. or 5 ft. These bolts should have an L-bend on the bottom and be mortared in fully so they don't pull out when the nut is tightened against the sill plate. They should extend about 2 in. out of the top of the wall. In some parts of the country, building codes require that the walls include a steel-reinforced, poured-in-place concrete bond beam in every fourth course.

Waterproofing the foundation—The traditional method of waterproofing a concrete block foundation is to parge (stucco) on two coats of mortar and then to apply a tar compound on top of that. This double protection works well, unless there is a severe drainage problem, and the soil is liable to hold a lot of water for a long time.

There are various mortar mixes you can use to parge the foundation. I recommend using a mix of one part portland cement to one-half part hydrated lime to three parts washed sand. This is a little richer than standard masonry cement and is known as type S mortar. The mix should be plastic or workable enough to trowel on the wall freely. Many mortars on the market that have waterproofers in them are all right to use. However, no two builders I know seem to agree on a mix, and most have worked out their own formulas.

Prepare the foundation wall for parging by scraping off mortar drips left on the block. Next, dampen the wall with a fine spray of water from a garden hose or a tank-type garden sprayer. Don't soak the wall, just moisten it. This prevents the parging mortar from drying

Photos: Dick Kreh

Troweling technique

Laying up concrete blocks with speed and precision takes a lot of practice. But it's chiefly a matter of learning several tricks, developing trowel skills and performing repetitive motions for several days. A journeyman mason can lay an average of 200 10-in. or 12-in. blocks in eight hours. A non-professional, working carefully and after practicing the techniques shown here, ought to be able to lay half that many. If you've never laid block before, what follows will show you the basic steps involved in laying up a block wall.

1. First, mix the mortar to the correct stiffness to support the weight of the block. Then apply mortar for the bed joints by picking up a trowelful from the mortarboard and setting it on the trowel with a downward jar of the wrist. Then swipe the mortar onto the outside edges of the top of the block with a quick downward motion, as shown.

2. Apply the mortar head joint pretty much the same way. Set the block on its end, pick up some mortar on the trowel, set it on the trowel with a downward jerk and then swipe it on the top edges of the block (both sides).

3. After buttering both edges of the block with mortar, press the inside edge of the mortar in the head joints down at an angle. This prevents the mortar from falling off when the block is picked up and laid in the wall.

4. Lay the block on the mortar bed close to the line, tapping down with the blade of the trowel until the block is level with the top edge of the line. Tap the block in the center so you won't chip and smear the face with mortar. Use a hammer if the block does not settle easily into place.

5. The mortar in the head joint should squeeze out to form a full joint at the edge if you've buttered it right. The face of the block should be laid about 1/16 in. back from the line to keep the wall from bowing out. You can judge this by eyeballing a little light between the line and the block.

6. Remove the excess mortar that's oozing out of the joint with the trowel held slightly at an angle so you don't smear the face of the block with mud. Return the excess mortar to the mortarboard.

Check the height of the blockwork by holding the story pole on the base and reading the figure to the top of the block. Courses should be increments of 8 in.

Finishing the joints—Different types of joint finishes can be achieved with different tools. The most popular by far is the concave or half-round joint, which you make by running the jointing tool through the head joints first, and then through the bed joints to form a straight, continuous horizontal joint. If you buy this jointing tool, be sure that you get a convex jointer. These are available in sled-runner type or in a smaller pocket size. I like the sled runner because it makes a straighter joint.

After the mortar has dried enough so it won't smear (about a half-hour), brush the joints lightly to remove any remaining particles of mortar. —D.K.

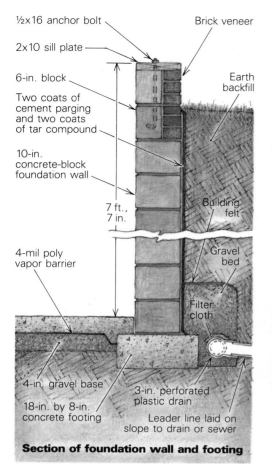

½x16 anchor bolt

Brick veneer

2x10 sill plate

6-in. block

Two coats of cement parging and two coats of tar compound

10-in. concrete-block foundation wall

Earth backfill

7 ft., 7 in.

Building felt

Gravel bed

4-mil poly vapor barrier

Filter cloth

4-in. gravel base

18-in. by 8-in. concrete footing

3-in. perforated plastic drain

Leader line laid on slope to drain or sewer

Section of foundation wall and footing

The completed foundation has been sealed with two ¼-in. thick parging coats and topped with an application of tar compound, which finishes the waterproofing. Backfilling should happen only after the first floor is framed and the walls framed up, so the added weight of the structure will stiffen the walls and make them less liable to bulge from the pressure of the earth.

out too quickly and allows it to cure slowly and create a better bond with the wall.

Start parging on the first coat from the bottom of the wall to the top, about ¼ in. thick. A plastering or cement-finishing trowel is excellent for this. After troweling on the parging, scratch the surface with an old broom or a tool made for this purpose. Let the mortar dry for about 24 hours or until the next day, and repeat the process for the second coat. Dampen the wall between coats for a good bond. Trowel the final coat smooth, and let it dry for another 24 hours.

To complete the waterproofing job, spread on two coats of tar compound (photo, top). You can do this with a brush or roller if the weather is warm. Many builders in my area use a product called Hydrocide 700B (Sonneborn Building Products, 57-46 Flushing Ave., Maspeth, N.Y. 11378). It comes in 5-gal. containers and is available from most building-supply dealers. I like it because it stays a little tacky and seals the wall very well. It's gooey, though, so wear old clothes and gloves when you're applying it. Kerosene will get it off your hands and tools when the job is done.

Drain tile—Most codes require some type of drain tile or pipe around the foundation to divert water build-up and to help keep the basement dry. The design of the drain-tile system is important. Generally, drain tile or pipe is installed around the exterior wall of the foundation, below the wall but above the bottom of the footing, as described below.

Begin by spreading a bed of crushed stone

or gravel around the foundation next to the wall. Lay the drain tile or perforated plastic pipe on top of this bed. The bottom of the drain pipe should never be lower than the bottom of the footing, or it won't work properly. Lay filter cloth over the drain pipe to keep mud and dirt from blocking the holes. Then place another 4-in. to 6-in. layer of crushed stone or gravel over the pipe, as shown in the drawing above.

The water collected by the drain tile has to flow away from the foundation. One way to make this happen is to drain the water to a natural drain away from the foundation area by installing a leader line on a slope that is lower than the drain pipe. The other method is to drain the water under the wall of the foundation and into a sump pit inside the basement. The water that collects is then pumped into a pipe up to grade or street level and allowed to drain away there naturally.

A third method, which has worked well for contractors in my area, is to put the drain tile inside the foundation on a bed of crushed stone, just beneath the finished concrete floor, which will be poured after the drain tile is in place. One-inch plastic pipe is installed about 6 ft. o.c. through the wall at the bottom of the head joints in the first course of block. When the foundation wall is done, crushed stone is spread around the exterior edges as before, but no drain pipes are needed.

The idea is that any water that builds up outside the foundation wall will drain through and into the drain tiles. In addition, the water inside the foundation area will also flow into

the drain tile and into the sump pit in the basement floor, where it can be pumped out to grade level to drain away, as mentioned above.

Insulating the foundation—In recent years, the use of rigid insulation applied to the exterior of the foundation wall has helped to reduce dampness and heat loss. This is especially important in the construction of earth-sheltered homes. There are a number of products that will do a good job. Generally, rigid insulation is applied to the waterproofed wall with a mastic adhesive that's spread on the back of the foam board. Use a mastic or caulking that does not have an asphalt base. Most panel adhesives will work. The building-supply dealer who sells the insulation will know the proper adhesive to use for a specific type of insulation board. Also, there are granular and other types of insulation that can be poured into the block cells.

After all of the foundation work has been completed, the backfilling of earth should be done with great care so that the walls don't get pushed out of plumb. It is always better to wait until the first floor is framed up before backfilling. This weight resting on the foundation helps prevent cracking of the walls, and the framing material will brace the block wall and make it more rigid. If the walls are cracked or pushed in from backfilling, the only cure is very expensive—excavate again and replace the walls. □

Dick Kreh is an author, mason and industrial-arts teacher in Frederick, Md.

Foundations on Hillside Sites

An engineer tells about pier and grade-beam foundations

by Ronald J. Barr

Most houses built on conventional sites sit atop a spread-footing foundation. It has a T-shaped cross section (small drawing, below center), and supports the house by transferring its own weight and the loading from above directly to the ground below. Spread footings can be used on sloping sites by stepping them up the hill (small drawing, below right), but this usually requires complicated formwork and expensive excavation. For slopes greater than 25°, the structurally superior pier and grade-beam foundation (large drawing, below) may be less costly. Apart from making steep sites buildable, pier and grade-beam foundations make it possible to build on level sites where soils are so expan-

sive that they could crack spread footings like breadsticks as the earth heaves and subsides.

How they work—The pier and grade-beam foundation supports a structure in one of two ways. First, the piers can bear directly on rock or soil that has been found to be competent. This means it's stable enough to act as a bearing surface for the base of the pier. Second, the piers rely on friction for support. This is what sets pier and grade-beam foundations apart from other systems. The sides of the concrete piers develop tremendous friction against the irregular walls of their holes. This resistance is enough to hold up a building. A structural engineer can look at the soil report

for a given site, study the friction-bearing characteristics of the earth and then specify the number, length, diameter and spacing of the piers that will be necessary to carry the proposed structure.

The rigs used to drill pier holes for residential foundations have to be able to bore holes 20 ft. deep or more. Foundations have to extend through unstable surface soil, such as uncompacted fill, topsoil with low bearing values and layers of slide-prone earth. Anchored in stable subterranean strata of earth or rock, deep piers don't depend on unstable surface soil for support.

The slope of the lot and the quality of the surface soil influence the size and placement

A pier and grade-beam foundation. Deep piers extend through layers of loose topsoil to lock into the stable soil below, which can support the weight of a house. The grade beams follow the irregular contour of the site, and transfer the building's loads to the piers.

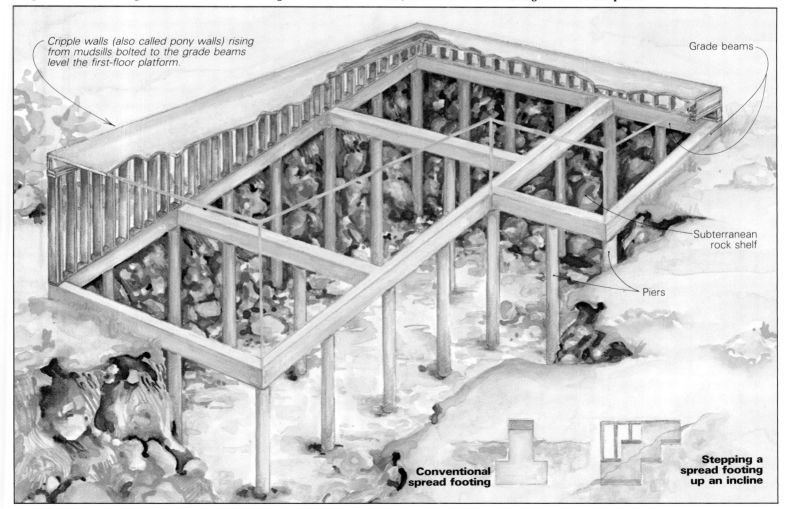

Cripple walls (also called pony walls) rising from mudsills bolted to the grade beams level the first-floor platform.

Grade beams

Subterranean rock shelf

Piers

Conventional spread footing

Stepping a spread footing up an incline

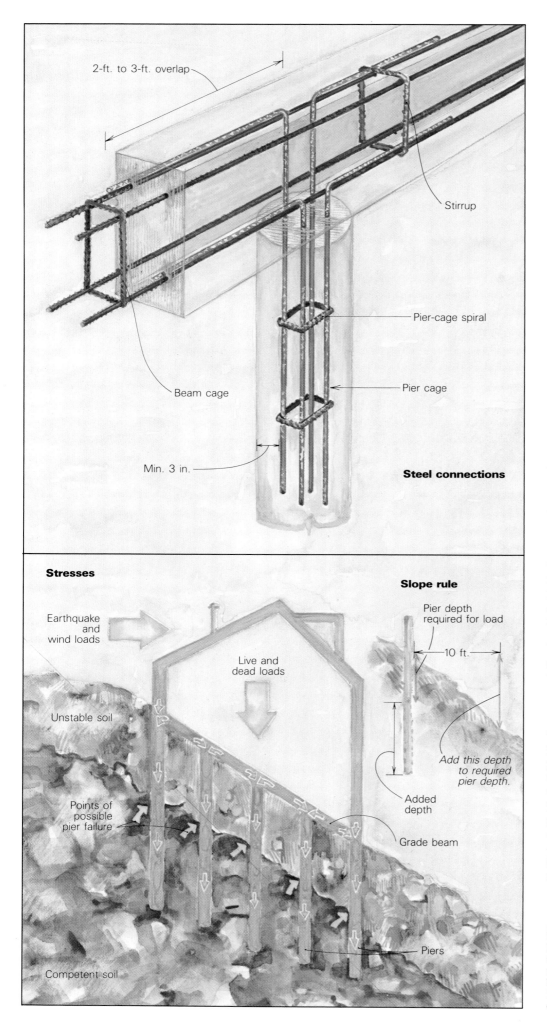

2-ft. to 3-ft. overlap

Stirrup

Pier-cage spiral

Beam cage

Pier cage

Min. 3 in.

Steel connections

Stresses

Slope rule

Earthquake and wind loads

Live and dead loads

Pier depth required for load

10 ft.

Unstable soil

Add this depth to required pier depth.

Points of possible pier failure

Added depth

Grade beam

Piers

Competent soil

---

of reinforcing steel in each pier (drawing, left). The steeper the lot and the more suspect the soil's stability, the bigger the steel. This is because the pier doesn't just transfer the compression loads from the house to the ground. It also resists the lateral loads induced by winds, earthquakes or by the movement of surface soils, which result in a kind of cantilever beam-action on the piers, as depicted in the drawing below left. They have to be strong enough to resist these forces—pier cages made from 1-in. rebar are not uncommon in extreme cases.

Grade beams—As the name implies, a grade beam is a concrete beam that conforms to the contour of the ground at grade level. Grade beams require at least two lengths of rebar, one at the top and one at the bottom. Foundation plans often call for two lengths at the top and two at the bottom, tied into cages with rectangular supports called stirrups. The bars inside the beams work like the chords in a truss, spreading the tension and compression forces induced by the live and dead loads of the structure above.

The sectional size of the grade beam depends on the load placed on it, and on the spacing of the piers. Pier spacing is governed by the allowable end-bearing value of the piers, or by the allowable skin friction generated by the pier walls and the soil under the site. Pier depth ranges from 5 ft. to 20 ft.; pier spacing, from 5 ft. to 12 ft. o.c. Pier diameters are normally 10 in. to 12 in., and can go up to 30 in. Beam sections start at about 6 in. by 12 in. and go up.

Because of the number of variables involved in designing a pier and grade-beam foundation, the engineering is best left to a professional. It's not happenstance that there are lots of lawsuits over foundation failures—many of them on slopes.

What the engineer needs—An engineer designing a pier and grade-beam foundation has to know what lies beneath the surface of the site. A soils engineer's report is usually required for this information (see p. 63). Sometimes the city or county building department will have records of soil characteristics in your area. The engineer also needs a copy of your house plans to calculate the loading on the foundation. If you can, tell the engineer what pier size your excavator can drill. If you're planning on making the steel cages yourself, ask your engineer about rebar diameter requirements. Although he may specify #5 or #6 rebar, which is impossible to bend in the field, it could be that more #4s (½-in. rebar) would do the job just as well.

Make sure you find out the spacing and size

Engineering considerations. **The steel embedded in each pier must resist the lateral loads from wind or earthquakes, which are transferred to the piers at the junction of the unstable and competent soils. If piers are embedded in a slope, the amount of slope rise in a 10-ft. run must be added to the pier depth required for load.**

Pier and grade-beam in expansive soils. A well-engineered system has a 2-in. gap between the grade beam and the soil, to allow for expansion. The pier should be straight, not bell-shaped at the top, so loads are transferred down to the stable soil.

of the anchor bolts your foundation will need for fastening sills to the grade beams. For some downhill slopes, the cripple walls (drawing, p. 51) will need additional blocking, and anchor bolts placed 2 ft. o.c.

Piers on a slope must be longer to compensate for the lack of soil on the downhill side (inset drawing, facing page, bottom). Figure the additional length as follows: Tie one end of a 10-ft. long string to the stake marking the pier location, and pull it taut downhill and hold it level. The distance from the end of the string to the ground directly below has to be added to the pier hole to compensate for its location on a slope.

Void boxes—Ask your engineer whether you can cast (pour) the grade beams directly on the soil, or whether they should have void boxes under them. A void box (drawing, right) is a gap, usually about 2 in., between a grade beam and expansive soils. Some builders put a 2-in. thick piece of Styrofoam at the bottom of the grade-beam forms before the pour. Once the forms have been stripped, they pour a solvent along the base of the grade beam to dissolve the Styrofoam. Another method is to fold up cardboard boxes until you've got a stack 2 in. thick, and then lay them at the bottom of the forms. You don't have to remove them—they'll rot away.

When I worked as a building inspector, I saw what can happen to a house when void boxes aren't included in grade-beam pours over expansive soils. A couple of home owners asked me to find out why their foundation had settled in the center of the house. There was a large fireplace there, and they thought it was the culprit. But it wasn't. The fireplace showed no signs of movement. However, outside the perimeter grade-beam foundation, recent rains had caused the clay to expand, raising the foundation (and the exterior walls) as much as 2 in. in places. The center of the house had stayed put, and everything else had risen around it. None of the doors or windows worked, and every wall in the house was cracked. Voids under the grade beams would have prevented all this.

Problems—Once the foundation design has been determined and your driller is on site, it's almost inevitable that something unexpected will happen. For instance, your excavator may not be able to drill to the depth specified on the plans. Sometimes you can get away with this; other times you have to provide for another pier nearby. If you're drilling friction-bearing piers, you may have to change their diameter and spacing, or change their end-bearing condition. Check with your engineer for the correct course of action.

Drill alignment sometimes slips during drilling, and the auger can break into an adja-

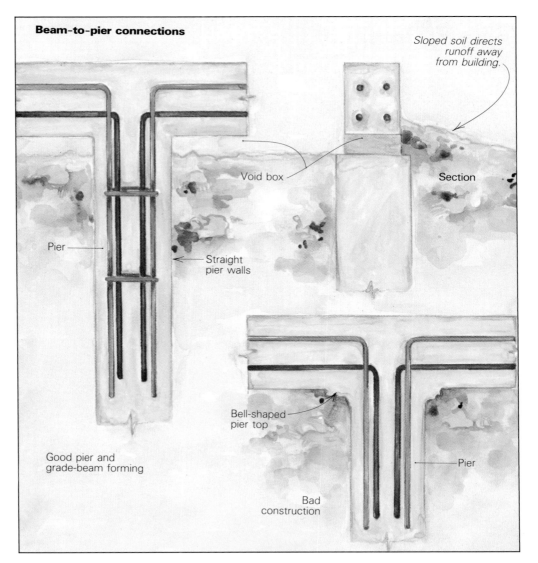

Beam-to-pier connections

Sloped soil directs runoff away from building.

Void box

Section

Pier

Straight pier walls

Good pier and grade-beam forming

Bell-shaped pier top

Pier

Bad construction

cent pier hole. If this happens, check with your structural engineer. The remedy will depend on the load-bearing requirements at that particular part of the foundation.

Watch the soil as it's being removed from the pier holes. Its appearance should change in the upper half of the hole, as the drill leaves the topsoil layer and passes into the more stable soil. If you notice an abrupt change in the color or texture of the soil at the bottom of the hole, get in touch with your soils engineer right away, because a change in soil type may mean you'll need to drill deeper piers to make sure you're into solid earth. Shallow piers are the cause of most pier and grade-beam failures, so it's best to err on the side of caution and drill those holes deep.

A little while ago I drew up some plans for remedial foundation work for a house that was suffering from short piers. Actually, some of the piers were long enough to extend into solid material, and this uphill portion of the house hadn't moved. But the downhill half rested on shallow piers that weren't engaged with the competent soil layer. The awesome power of the surface layers moving downhill had pulled the house's grade beams apart and stretched their reinforcing rods like taffy. Above these cracks in the beams, the house was slowly being torn in half. Just getting a

drill rig to the site required dismantling part of the house, and the necessary foundation repairs ended up costing $60,000.

After the drilling—Grade beams should be centered over the piers. If one is out of alignment by more than a few inches, the eccentric load placed on the pier could eventually crack it in half. Rebar splices should overlap by 3 ft. And make sure that the tops of the pier holes haven't been widened to create a bell shape (drawing, above). Such a bell can be a bearing surface for expansive soils, and can cause uplift problems during wet weather.

A pier and grade-beam foundation usually doesn't need a special drainage system, but it's important to direct runoff away from the building. Banking the soil a few inches up the exterior side of the grade beam is generally all that's needed, and won't appreciably intrude into the void under the beam (drawing, top). The important thing is to make sure that running water doesn't wash away the soil around the piers. This erosion might not be critical if the piers are end-bearing, but if they are friction piers, the washing action could severely limit their load-carrying capabilities. □

Ronald J. Barr is a civil engineer. He works in Walnut Creek, Calif.

Foundation Drainage
Retrofitting is doing it the hard way

by David Benaroya Helfant

In a hillside community of homes I inspected recently, I found an interior stairway that rests on a "floating" concrete slab-on-grade. The owners of the house complained about the constant repairs needed to patch the drywall and baseboards that were connected to this stair. It seems that every time it rained, the stair would move up relative to the rest of the house. When the ground dried out, the stair sank.

The problem was that the stemwall foundation around the house was poorly drained, resulting in an inconsistent moisture content in the soil under the footings and allowing runoff to migrate beneath the slab some distance from the exterior stemwall. The slab was placed atop the undrained clay-laden soil, and there was negligible weight on the slab. Every time it rained, the clay expanded, taking the stairway for a short ride upward, tearing the drywall joints and wracking the baseboards, railings and casings out of alignment. The

remedy to the situation was the placement of a drain uphill from the slab that diverted the water away from it.

The objective in designing and installing drainage systems around the perimeter of a foundation is to keep water from soaking into the soil and moving under the footings. Water initiates the undermining of a foundation by causing erosion beneath it, literally carrying away the soil upon which the footing bears. Often this is the prelude to building settlement. If the soils have a high clay content, poorly drained foundations can be cracked or rotated by forces exterted by the wet clay as it expands (photo above).

If the grade around a house is well-sloped, you may not need a subsurface drainage system. But if your house is on a hillside made up of soils that drain poorly, such as clay, subsurface drains can be essential to the long-term well-being of the structure.

In most cases, a structure won't be threatened with a terminal illness brought on by bad drainage, but it can suffer an abundance of minor maladies. A damp crawl space can cause mustiness, mold and mildew in a house, and fungus wood rot and termites thrive in this kind of environment. Soils under foundations that undergo dry/soggy cycles can bring on the familiar phenomenon of sticking doors and windows. While good site drainage may not solve all moisture problems (such as condensation), it can be effective in combatting cyclical changes, such as the floating-slab phenomenon described before.

System basics—An effective drainage system consists of two distinct systems: a subsurface drain to carry away the flow of ground, or subsurface water, and a surface drain to convey rain or snowmelt away from the building (drawing, facing page). The core of any subsurface drainage system is a network of perforated pipes laid at the bottom of trenches next to or near the foundations, and sloped to drain toward a suitable receptacle. The pipe is laid with the perforations pointing down, so that water seeping into the trench from below will rise into the pipe and be carried off. Above the perforated pipe is a run of unperforated pipe (photo right) that is used to transport runoff from roofs, patios, walkways and other paved surfaces. Typically, these pipes will lead to a dumping site 10 ft. to 20 ft. downhill from the house (more on this later).

You may ask, "Why can't I just run my downspouts into the subsurface drain, and do away with the surface drain?" Don't do it. Combining the two increases the potential for a clogged line, and it defeats the purpose of a subdrain by injecting water into the ground.

For most residential drain lines—both surface and subsurface—we use 3-in. dia. pipes. But if we're working on a hillside where we expect heavy flows, we'll install 4-in. subsurface lines. If a large roof area is draining into a single downspout, we'll play it safe and install a 4-in. surface drain line to carry the runoff.

Pipes and fittings for drain lines are quite similar to those used for DWV (drain, waste and vent) work, but the fittings don't come in as many configurations and the pipes aren't as heavy. They also cost less—about 40% to 50% of what corresponding DWV materials cost.

We prefer to use smooth-wall pipe and fittings made of polyethylene plastic for subsurface and surface lines. This is a fairly rigid pipe that can be cleaned by an electric snake without being diced up from the inside out. We specify pipe that is rated at 2,000 lb. of crushing weight. This is important because subsurface pipes are often buried well beneath the surface, and we usually compact the earth above them. Where the line passes under a sidewalk or a portion of a driveway that will carry traffic, we switch to cast-iron pipe and link the two materials with no-hub couplings.

When we can't get the polyethylene pipe, we use polystyrene pipe instead. But the crew doesn't like to work with this material because it's brittle, which makes it tougher to assemble the fittings and pipe sections. We never use clay-tile pipe, which comes in 12-in. to 16-in. sections that butt against each other. There are too many opportunities for sections to move differentially. Nor do we use the thin, corrugated polyethylene pipe, as an electric snake can rip it apart if a line needs augering to clear a blockage. Its weak walls make it suspect for deep trenches, and its fittings do not seem to seal well.

In deep systems where it is necessary to carry large volumes of water, corrugated galvanized-steel or thick-walled ABS or PVC pipe may be preferable. Under these conditions, you should consult a geotechnical engineer for specific recommendations regarding dimensions and types of pipe.

All of our drainage systems are designed with cleanouts, similar to conventional wasteline plumbing systems, so that the system can

Two pipes. To properly drain a structure, you need to pick up subsurface water moving through the soil as well as runoff carried from flat surfaces and roofs. The pipe in the bottom of the trench above is perforated, with its holes oriented downward. Gravel covers it to within a foot of grade, and filter fabric is folded over the gravel to keep fines from clogging the drain line. The top pipe carries surface runoff from downspouts and drains. The pipe stub projecting above the tamped earth will attach to an area drain.

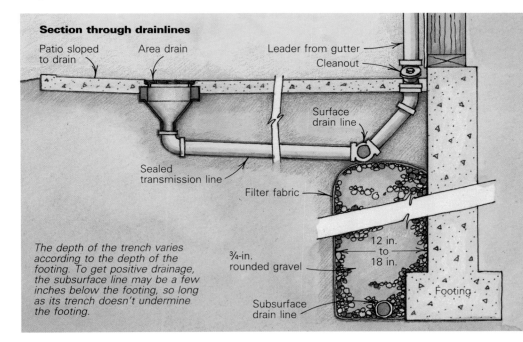

Section through drainlines

Patio sloped to drain · Area drain · Leader from gutter · Cleanout · Surface drain line · Sealed transmission line · Filter fabric · ¾-in. rounded gravel · 12 in. to 18 in. · Footing · Subsurface drain line

The depth of the trench varies according to the depth of the footing. To get positive drainage, the subsurface line may be a few inches below the footing, so long as its trench doesn't undermine the footing.

be cleaned with an electric snake. We put them at 30-ft. to 40-ft. intervals on straight runs and at strategic locations elsewhere: major bends, intersections with other lines and the point at which downspouts enter the surface-water drain system (a prime target for a leaf clog). We always cap the cleanout with a plug so that it does not collect debris.

We handle surface-drain flow either with an area drain, a catch basin, or a trench drain (drawing, p. 57). The first two are concrete, alloy or plastic boxes that have metal grills to keep debris out of the systems. The area drain is connected to a leader that ties into the surface drain line (drawing above). A catch basin collects water from several surface drains and feeds it into a single outlet. A catch basin is also deep enough to allow sand and soil to fall to the bottom, where they collect without interrupting the water flow. These "fines" settle into

Retrofit drainage work begins with excavation of a 12-in. to 18-in. wide trench to the base of the footing (photo above). You know you've got a subsurface water problem when your basement is 9 ft. below grade, water is 5 ft. below grade and the remains of your drain line are 1 ft. below grade. It can be seen in the center of the photo (left), on the trench's right bank. At the high end of the system (photo below), cleanouts should be installed allowing access in both directions. Shown in this photo is the subsurface line—the surface line will also need cleanouts.

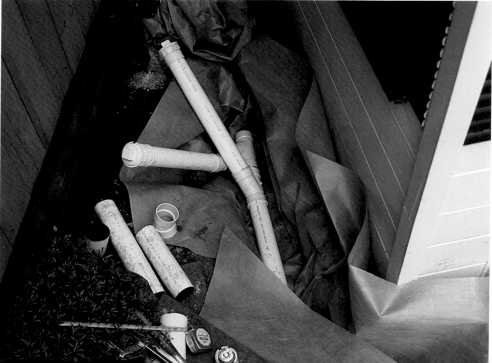

a sludge that should be removed periodically. Trench drains are long, narrow steel, plastic or fiberglass boxes with grills on them. You see them at the base of driveways, where they catch the water before it inundates a garage that's downhill from the street. We usually install the ones made by Polydrain (ABT, Inc., P. O. Box 837, Murdock Rd., Troutman, N. C. 28166).

Tools of the drainage trade are neither mysterious nor high-tech. They include picks and shovels as well as pneumatic and electrical demolition and digging tools to break up the earth. Good wheelbarrows are essential. We recently acquired a Takeuchi tractor, which has a backhoe and an auger attachment. It speeds up excavation considerably, but in some cases, even a little tractor like this is tough to maneuver. Therefore, drainage work on steep hillsides tends to be labor-intensive, and impossible without a conscientious crew.

Putting it in the ground—Naturally it's best to think about controlling subsurface water in the planning phases of a new construction project. The pipes can be installed alongside the new footings before the trenches are backfilled. But I can assure you, plenty of houses have been built with inadequate systems for controlling ground water, if any at all. The methods for assembling a system during new construction are the same as that for a retrofit—it's just a lot easier to do it before the foundation trenches are backfilled and the landscaping has taken root. We find that it costs two to three times more to install a drainage system around an established house than it does to add drainage around a new house. The photos illustrating this article show a couple of typical retrofit installations.

When we can get the tractor into place, we begin a job by trenching alongside the foundation (top photo, facing page) until we've reached the base of the footing without undermining it. In the project shown in the photos on p. 54, facing page above and below left, the house was cut into the hillside to create space for a garage, and the cast-concrete foundation and the stairs to the side yard were gradually being shoved east by the swelling of the clay-laden soils. When this house was built in the '20s, the builders had included foundation drain lines. But they were cast-iron pipes installed a foot or so below grade. When we found them, they were rusted out, clogged with mud and totally useless. We also found lines as we trenched along the back of the house (bottom left photo, facing page) and hit water at about 5 ft.—4 ft. above the level of the garage slab.

Before we laid pipe in our trenches, we lined the trenches with geotextile fabric. Also known as filter fabric, this material is made of either woven or spun-bonded polyester or polypropylene fibers. The purpose of the fabric is to keep the fines in the soil from migrating into the gravel backfill, which eventually would clog the drain line. We use a 4-oz.-per-yard, spun-bonded fabric designed for soils that don't have a lot of silt in them. Under some conditions, soils that have high sand and silt contents can clog filter fabrics, so if you're in doubt about the makeup of the soil, have a soils lab test its constituents so you can choose a fabric accordingly. Filter fabric is remarkably tough stuff. To cut it we use sharp sheet-metal shears, but a sharp razor knife will work. We pay about 15 cents per sq. ft. for the fabric and buy it from a local vendor that supplies products related to concrete work. If you can't find filter fabric locally, two companies that make it are Mirafi (P. O. Box 240967, Charlotte, N. C. 28224) and Hoechst Celanese Corp. (Spundbond Division, P. O. Box 5887, Spartanburg, S. C. 29304).

Before the fabric is down, we sometimes add a thin layer of sand to smooth out the bottom of the trench. This keeps the pipes from getting flattened at the high spots when the gravel backfill is placed. But if the bottom of the trench is pretty uniform, we skip the sand.

The pipes have to be sloped at a minimum of ⅛ in. per ft. For shallow trenches that have relatively short runs—say 40 ft.—we'll use a 4-ft. level with gradations on the bubble that read slopes of ⅛ in., ¼ in. and ⅜ in. For longer runs, or deeper trenches, we use a transit and a rod to check the slope.

We cut pipes with a hacksaw, and for the most part, we rely on press-fitting the parts because they usually go together with a satisfying snugness. If so, we don't bother gluing them. But if they seem loose, we swab them with the glue supplied by our vendor for the particular kind of pipe and wrap them with duct tape as a further hedge against separation during backfilling.

When all the subsurface lines are in place and their cleanouts have been extended above grade, we fill the trench with gravel to within a foot of grade. This gravel should be clean ¾-in. material. If you are applying polyethylene sheeting to the foundation as a moisture barrier, use rounded rock. Crushed rock has sharp edges that will damage the poly. Otherwise, crushed rock is usable and probably cheaper. Don't use road-bed mix, though, because it has too many fines in it. Gravel in place, we wrap the fabric over the top like a big burrito.

Atop all this goes the unperforated surface runoff lines (photo, p. 55). While we usually position them about a foot below grade, they can be placed lower, if necessary, for positive drainage or to protect them from the gardener's shovel. The surface runoff lines, too, are sloped at least ⅛ in. to a foot. Leaders

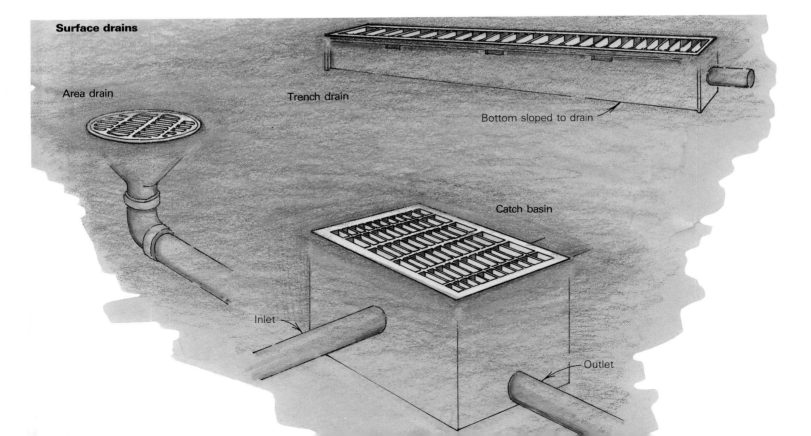

Surface drains

Area drain

Trench drain

Bottom sloped to drain

Catch basin

Inlet

Outlet

from the rain gutters are connected to the surface lines by way of plastic fittings that are square on one end to accept the leader, and round on the other. At each leader entry we place a wye fitting for a cleanout, and at the highest elevation of the surface drain line we position a pair of cleanouts (photo, p. 56).

Usually the surface line and the subsurface line will drain to daylight at different places, but we sometimes combine the two if we need to go under a sidewalk with a line. In this case we make sure the intersection is at least 20 ft. downhill from the structure to minimize the chance of a blockage that would cause water to back up into the subsurface line. And we include a cleanout.

To finish the job, we will either compact a layer of soil on top of the buried lines, or cover them with more gravel if the surface is likely to get heavy runoff. Road base is usable here. It's probably overkill, but I think it's best to top the system with some type of paving—either a poured-concrete cap or individual pavers that direct the water away from the building.

What to do with the water— The final phase in drainage work is doing something with all that water once you've got a system for collecting and rerouting it. Two guidelines are important to follow. First, if bad drainage is causing your property to deteriorate, then it's important to make sure that your depository doesn't cause the same problems, albeit in a different location. Second, don't put your storm drainage on your neighbor's property. In the latter case when you're on a hill, sometimes the only appropriate solution is to secure an easement for a drain line that will discharge below both properties.

In many places storm drainage must be put into established sewer systems dedicated to carrying runoff. Local jurisdictions vary on the hookups required. In some cities you simply daylight the drain lines at the curb, sending the water to the storm sewers via the gutters, while other jurisdictions require a sealed hookup. Some will let you dump runoff into effluent sewer lines, but I've found this to be the exception. Given the differing approaches, make sure you check with your local building department to verify local practices.

If there isn't a municipal storm sewer to carry away the water that is collected, things get more complicated. Leach fields, energy-dissipation basins and dry wells are three approaches to allowing the runoff to continue draining slowly, in a manner less likely to cause erosion and other related problems. But these strategies can concentrate an abnormal amount of water in one place. So if you're unsure about the stability of the soils at a likely water-dump site, you should consult a soils engineer.

If you have an open area with suitable soil, a leach field can be used to distribute the water back into the ground. Like a septic leach field, this requires a manifold of perforated pipes buried below grade. You run a

sealed transmission line into one end, and the water will dribble back into the soil over a large area.

An energy-dissipation basin takes less space than a leach field. It usually consists of an excavation lined with filter fabric and filled with rock graded by size (drawing below). The ones that we've done have been about 5 ft. square and 4 ft. deep. The graded rock is arranged so that the big ones are on the bottom where they can get a good bite into the hillside. A sealed

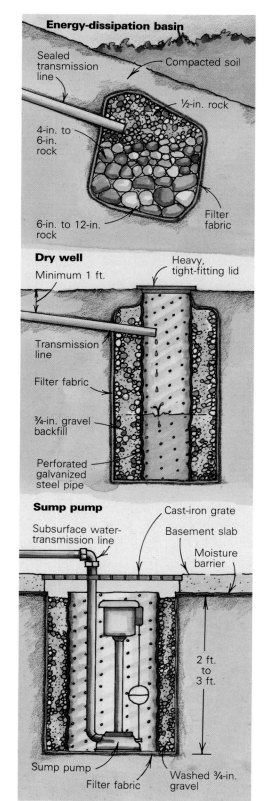

Energy-dissipation basin

Sealed transmission line
Compacted soil
½-in. rock
4-in. to 6-in. rock
6-in. to 12-in. rock
Filter fabric

Dry well

Minimum 1 ft.
Heavy, tight-fitting lid
Transmission line
Filter fabric
¾-in. gravel backfill
Perforated galvanized steel pipe

Sump pump

Cast-iron grate
Subsurface water-transmission line
Basement slab
Moisture barrier
2 ft. to 3 ft.
Sump pump
Filter fabric
Washed ¾-in. gravel

line enters the basin, and its end should be buried in about 2 cu. ft. of ½-in. rock. Recently we had to convey a load of 4-in. to 12-in. dia. rocks 200 ft. down a hillside. We made staging areas at 60-ft. intervals and rolled rocks to them through taped-together Sonotubes. The crew loved it.

Covered with a layer of native soil, an energy-dissipation basin can be made that will virtually blend into the landscape. But make sure you compact the soil before landscaping to avoid the inevitable settling that will occur.

If there is insufficient grade for positive drainage away from the building, a dry well (called a *stand pipe* in the midwest) could be your solution (see drawing). It is usually placed 15 ft. to 20 ft. from the house. When constructing one of these, we use an 18-in. to 24-in. dia. corrugated and perforated galvanized-steel pipe set vertically into an excavation roughly 4 ft. in dia. and 8 ft. to 10 ft. deep. The hole is lined with filter fabric, the pipe goes in the middle and a collar of ¾-in. drain rock fills the space between them. A sealed transmission line from the drainage system, sloped at ¼ in. per ft., enters the dry well. It should be at least 1 ft. deep to protect it from shovels and roto-tillers.

A dry well should be capped with a heavy, tight-fitting lid that is impervious to youngsters. I hasten to add, however, that dry wells are not appropriate alternatives in all situations. Installed on slopes, they may concentrate water where you don't want it, and could even create an unstable slope condition that might result in a landslide.

As a last resort, you can use a sump pump (see drawing) to lift water out of an undrainable situation. The sump pump goes in the deepest part of the basement and needs a reliable source of power. Two essential characteristics make sump-pump systems less than ideal solutions. For one, if they are installed within the home's footprint, water is still getting in or under the building. And two, they rely on manufactured energy vulnerable to outages that typically occur precisely when you need the pump the most. But in some cases, a sump pump is the only way to get the water to an acceptable distribution system. Regardless of the system you elect to install, it's a recommended practice to keep good records showing the details (size, depth, location, cleanouts) of the system. □

David Benaroya Helfant, Ph.D., is managing officer of Bay Area Structural, Inc., general engineering contractors in Oakland, Calif., and is principal of Independent General Engineering Inspection Services, Emeryville, Calif. For more on drainage systems, see Soil Mechanics in Engineering Practice *by Terzaghi and Peck (Wiley and Sons, New York, currently out of print),* The Earth Manual *by Malcolm Margolin (Heyday Books, P. O. Box 9145, Berkeley, Calif., 94709, 1985. $12.95; 252 pp.) and* Roadside Geology of Northern California *by Alt and Hyndman (Mountain Press Publishing, Box 2399, Missoula, Mont., 59806, 1975. $9.95; 245 pp.).*

Insulating and Parging Foundations
Covering concrete walls with rigid foam insulation and troweling on stucco requires experience with the materials

by Bob Syvanen

If you've got the idea that a builder's skill is an unchanging body of knowledge passed down through the generations, think for a minute about insulating a foundation from the exterior. Even in cold climates, what you used to see between the bottom of the siding and the grade was the bare concrete foundation wall. But these days, with estimates of heat lost in a house through the foundation running as high as 30%, what looks like concrete is more likely parging, or stucco, applied over rigid foam insulation.

Insulating the outside of foundations has been a problem for a lot of builders, including me, because many of the materials and methods are new. Although rigid foam-board insulation doesn't look like much of a problem, it isn't as simple as it first appears. Polystyrene is the insulating material most often used. It comes in 2-ft. wide panels and handles like plywood, but it's a lot lighter. You can cut it with anything from a knife to a table saw. But polystyrene foam is produced in two forms: expanded and extruded. Expanded polystyrene (EPS), also known as beadboard, is more susceptible to soaking up moisture than its extruded cousin, say the researchers on one side of this controversy. This could lead to a considerable loss in R-value. Although I used expanded polystyrene on the job shown here, I think the extruded version is probably the better bet despite its higher cost.

There are more than 100 makers of EPS; but extruded polystyrene is made in the U. S. by only three companies: Dow Chemical (Midland, Mich. 48640), whose blue-tinted Styrofoam is often called blueboard; Minnesota Diversified Products (1901 13th St. N. E., New Brighton, Minn. 55112), whose yellow product is trademarked Certifoam; and U. S. Gypsum (101 S. Wacker Dr., Chicago, Ill. 60606), the makers of pink Foamular.

Whichever brand you use, the process of applying it is the same, and so are the problems. For instance, asphalt-based products dissolve most foams, so my usual method of waterproofing a foundation is suddenly out

the window. And to complicate things even more, polystyrene needs to be protected above grade from impact as well as from deterioration by ultraviolet light from the sun.

I wanted a protective coating that was easy to install, good looking and long lasting. There are many commercial systems—fiberglass panels, super stucco mixes, and even a rigid insulation with a factory-applied coating that can be attached to concrete forms before pouring—but I wanted to use materials that were more traditional.

I first used asbestos board cemented on the foam, but it is fragile, hard to repair, impossible to glue, and required a lot of fitting time at corners, doors and windows. I also tried a latex-cement product applied directly on the foam. Unfortunately, it didn't age well. In fact, I have repaired not only the job I did with it, but several others in my area.

I finally settled on covering the foam with cement-stucco, called parging where I live. When stucco is used for exterior wall finish on a house, it is usually done in three coats like plaster. I was determined to come up with a single application process. Although parging and surface-bonding mixes can be applied directly to the insulation, I don't trust the bond, and want a thicker parging for durability. This means using some kind of lath.

On the first parging job, I used small-mesh chicken wire. I stretched it over ⅜-in. wood lath at 12 in. o. c. both horizontally and vertically to hold it off the surface of the insulation. Chicken wire wasn't the answer. Although my mason got the chicken wire to support the cement out of sheer stubbornness, the diamond-shaped pattern showed through a little, and there were some shrinkage cracks.

I refined the system by using metal lath,

and by reducing the thickness of the wood lath to ¼ in. This worked much better for the mason, but the lath strips were still tedious to install. Next I eliminated the wood lath and applied the metal lath directly to the foam. What I ended up with is a protective coating that is long lasting, attractive and relatively easy to install. Although it is a little expensive, after seeing some of the jobs using cheaper materials other local builders and I have done, I think it's worth the cost.

Since the insulation and lath-work usually fall to the carpenter or contractor who is on the site every day, the only sub I use is my mason, who is much faster and neater than I am with a trowel. I am used to paying anywhere from $2 to $4 a square foot for parging, although conditions vary enough that both the mason and I get the best deal when I use him on a time-and-materials basis. Since a bag of masonry cement covers 20 to 30 sq. ft. of wall, most of the expense is in the labor.

Installing rigid foam insulation—A partially earth-sheltered, passive-solar house I just completed gave me a good chance to try my new system. The plans called for its concrete walls to be insulated with two layers of 2-in. foam. One wall is 7 ft. 10 in. high, and the other three are 2 ft. high. The parging was to cover the first 2 ft. below the mudsill on all of them. Some folks also use insulation laid horizontally below grade, but I simply ran my panels down to the footings.

The first thing I needed was a good adhesive, since there shouldn't be any give in the plane of the insulation panels if the parging is going to last. But the high wall is also below grade and part of the living space, so it had to be well waterproofed. Since asphalt-based products can't be used with foam, I looked

Installing foam insulation. First, a waterproofing agent that also serves as a mastic is spread directly on the concrete. Temporary braces hold the foam panels in place while the mastic dries. Two-by-four nailers are used between the two 2-in. layers. The horizontal nailer is 27 in. down from the sill—the width of the metal lath that will be applied next.

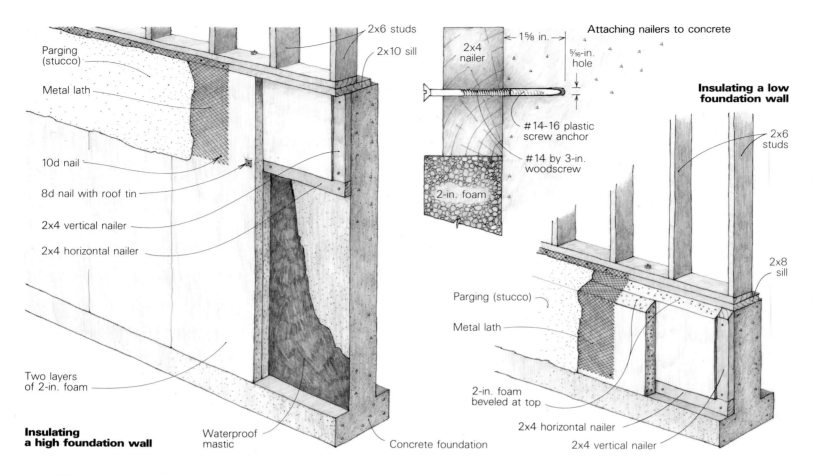

Parging (stucco)

Metal lath

10d nail

8d nail with roof tin

2x4 vertical nailer

2x4 horizontal nailer

Two layers of 2-in. foam

Insulating a high foundation wall

Waterproof mastic

2x6 studs

2x10 sill

Concrete foundation

2x4 nailer

← 1⅝ in. →

⁵⁄₁₆-in. hole

Attaching nailers to concrete

#14-16 plastic screw anchor

#14 by 3-in. woodscrew

2-in. foam

Insulating a low foundation wall

2x6 studs

2x8 sill

Parging (stucco)

Metal lath

2-in. foam beveled at top

2x4 horizontal nailer

2x4 vertical nailer

around for something else. What I found was a mastic, Karnak 920 (Karnak Chemical Corp. 330 Central Ave., Clark, N.J. 07066), which is marketed as both an adhesive and a water-proofing agent. Theoretically, you trowel the mastic waterproofing on the concrete and then press the foam panels in place. But the walls had enough irregularity that the panels contacted the mastic in only a few places, and they fell off about as fast as I put them on. I then found out that the foam has to be applied before the mastic skins over. This is enough time to apply just one or two panels and brace them with sticks, 2x4s, stones or buckets (photo previous page). When the foam was applied in this way, the adhesive held.

In this case, I installed 2x4 pressure-treated wood nailers with the first layer of foam panels in order to get nailing for the second layer. On the 7-ft. 10-in. wall, I began by placing a horizontal nailer 27 in. down from the sill. It was used to attach both the second layer of insulation and the lath, which comes 27 in. wide. Next, I attached vertical nailers above the horizontal at 24 in. o. c. because the panels are 2 ft. wide. I also filled in with nailers at windows and corners to catch the edges of the panels. A quick, easy way to fasten these 2x4s to concrete is to hold the nailer in place and drill through the wood into the concrete. Using a hammer-drill makes this almost fun.

For 2x4s, I use a #14-16 screw anchor, 1½ in. long, with a #14 by 3-in. flat-head woodscrew. Use a piece of tape on a ⁵⁄₁₆-in. masonry bit, at 3 in. from the tip, to limit the hole depth. Most hammer-drills have an at-tached depth guide. If the hole is too shallow, the tip of the woodscrew won't hit the con-

crete before snugging up the 2x4. If the hole is too deep, the screw won't grab the anchor.

After drilling, insert a plastic screw anchor into the hole in the nailer and turn a wood-screw a few turns into it. Then hammer the screw-and-anchor combination through the nailer into the concrete. Last, screw the wood-screw home (drawing, above center).

On the 2-ft. wall, I cut the first layer of foam to fit between the footing and the horizontal nailer and installed it before the nailers. This way, I could wedge the nailers between the foam and the footing while I fastened them.

On the high wall (drawing, above left), I cantilevered the mudsill over the concrete by the depth of the foam so that I could nail the metal lath directly to its top edge. This meant that the second vertical layer of foam tucked up underneath it, flush with its outside edge. This layer is held in place with 8d nails wher-ever there are nailers. The 8d nail reaches through to the nailer, and when given an extra tap, the foam compresses and snugs up the panel nicely. I use a roofing tin on each of these nails to increase its bearing surface. This is a stamped 2-in. by 2-in. flat metal plate with a hole in the center, and is typically used to hold down roofing felt on windy days. You can buy them from a roofing-supply yard or make your own by cutting out sheet-metal squares. I have also seen pins and plastic shields manufactured for this purpose.

The 2-ft. wall was insulated in a similar manner, but here the sill is flush with the out-side face of the foundation wall (drawing above right). Since the foam projects past the sill, I beveled the top edge of the foam at a 45° angle. The vertical 2x4 nailers were also bev-

eled at the top before I fastened them to the concrete. The corner nailers are beveled from each direction (photo facing page, top left). To make a neat bevel cut in the foam, I snapped a chalkline the length of the founda-tion on the face of the panels and sawed along it with a bread knife. This bevel design worked well here because the finished grade was to come at the bottom of the bevel.

Installing metal lath—The galvanized metal lath I use measures 27 in. by 96 in. It is sold in single sheets or in bundles of ten. There are two things to keep in mind when you're work-ing with metal lath. First, the diamond-mesh pattern is formed on an angle between the front and back of a sheet. This means that the dividing wire that is roughly horizontal forms a small lip or cup at the bottom of each hole. Make sure these cups are facing up to catch and hold the parging. It will work both ways, but things go better if the cups are up. The other thing to remember is that metal lath is sharp. I don't think I have ever worked with the stuff without cutting myself. The cuts are not bad, just annoying. Wearing gloves helps, but I find that more annoying than the cuts.

On the high wall, I nailed the mesh to the top edge of the sill. The bottom of the mesh nails through the foam into the horizontal nailer 27 in. below. I used leftover 3d shingle nails at the top, and 10d commons on the bot-tom and along the edges wherever I had a nailer. The 2-ft. walls were fastened similarly, but because of the beveled top, I had to bend the lath before I nailed it in place.

For corners, expanded corner bead—the plasterer's version of a metal sheetrock cor-

Photos: Bob Syvanen; Illustration: Frances Ashforth

ner—is the best way to go because it forms a neat, stiff straight line. But I didn't have any on the job, so I pre-bent the lath at 90° before installing it (photo top right). Bending sheet metal, particularly metal lath, on the job site isn't hard if you think of how it's done in the shop and duplicate the procedure. The shop uses a brake, which is a cast-iron table and a bar that folds the sheet metal over the edge of the table. On site, I sandwich the sheet between two 2x boards and "break," or fold, the piece that sticks out over the bottom 2x using a scrap block about 2 ft. long. Nailing the sheet metal to the top of the bottom 2x keeps it from creeping out as the bend progresses. This system is particularly good for metal lath because you don't have to handle the material constantly as you bend it.

Parging—Parging is not impossible for a novice, but a good finish takes experience. The first job I did turned out okay, but there was lots of room for improvement, so I went to school by watching mason John Hilley.

The parging he uses is a one-coat stucco with a steel-trowel finish. Other finishes might work better, but I am satisfied with this one. The mix he uses is 16 shovels of sand per bag of masonry cement. He doesn't have any trouble using up a batch that size before it begins to set. Masonry cement is a mix of portland cement, hydrated lime and additives that combine with water and sand to form mortar or stucco. For a parging mix, use Type M for higher compressive strength and greater resistance to water.

Large expanses of stucco are usually worked with darbies and floats. For foundations, though, a standard mason's trowel is easier. The mud is picked up on the bottom surface of the trowel and immediately applied to the lath. The free hand assists by pushing against the top face of the trowel, forcing the mud into the mesh. It is a quick process—pick up, apply, press. With each pressing motion, the excess cement gets pulled along with the sliding trowel (photo center right).

At the same time that the parging is applied, it should be roughly surfaced, to establish·an even thickness. As with brick jointing and slab work, compressing the material is what finishing is all about. This requires a bit of pressure, but it should be with good control. Use two hands on the trowel, one on the handle and the other on the flat of the blade, and keep your arms straight.

The finishing is done when the shine leaves the surface of the parging. The trowel is dipped into water, shaken once to get the excess off, then pressed against the surface of the stucco using both hands (photo bottom right). Try to get a smooth finish in just a few strokes so you don't overwork the cement.

A cloudy, cool day is best for parging because the mix can be worked longer before it sets. If the parged wall is in direct sun, mist the surface with a pump-up garden sprayer filled with water to keep the surface of the stucco from drying out too fast, which will cause shrinking and surface cracks.   □

Preparation. This foundation corner (top left) is ready for lath. Cutting the double bevel on the top of the corner nailer to match the bevel on the two layers of foam requires much less work later when the stucco is finish-troweled. An 8d nail and its roof tin, which acts like a large nailhead, are just visible at the bottom of the photo. This same corner is ready for parging once it is wrapped in metal lath (top right), which is pre-bent on a brake. The lath is nailed to the sill at the top, and through the second layer of polystyrene into the horizontal 2x4 nailer at the bottom.

Parging. Mason John Hilley forces the stucco mud into the lath (above), using two hands and the weight of his body. Just one coat of parging is used, but it is troweled twice. The first time is a rough troweling. When the shine disappears, the surface is smoothed with the same trowel, dipped frequently in water. This finish process is also done with one hand on the face of the trowel for direct pressure (right), and with arms held straight for good control.

A Pier and Grade-Beam Foundation

Advice from a contractor on one way to build on a steep slope

by Michael Spexarth

As good home sites get harder to find, many builders are looking at hillside lots that used to be considered unbuildable. A builder who can cope with the special foundations required on steep sites can often buy a lot for a modest sum, and build a house that enjoys an attractive view of the scenery below.

Here in the San Francisco Bay area, chances are good that you'll find a pier and grade-beam foundation under just about every new hillside home. This type of foundation links a poured concrete perimeter footing to the ground with a matrix of grade beams and concrete pilings, some up to 20 ft. deep. The resulting grid grips the hillside like the roots of a giant tree (see pp. 51-53). Slopes in excess of 45° can be built on with this kind of foundation. And in areas where there are landslides, expansive soils or earthquakes, a pier and grade-beam foundation may be required by local building codes.

As a rule pier and grade-beam foundations need more reinforcing steel than conventional foundations, and require special concrete mixes. On the other hand, they usually call for less formwork than perimeter foundations do. What's more, the footings for grade-beam foundations don't have to be level.

Soil survey—A soil engineer's report is usually required by local codes, and even if it isn't, it's smart to get one. Because the tendency on steep sites is to overbuild, the survey's recommendations may reduce the number and size of piers, which designers or engineers with less knowledge of geology will spec to err on the side of caution. This kind of needless overbuilding wastes money.

To make the survey, the soil engineer will make test drillings or trenches at various spots around the site with a drill rig that typically bores a 6-in. dia. hole. Or he will bring in a backhoe. At various depths, core samples are taken with a cylinder that is lowered into the test hole. Although they will vary from site to site, these holes are usually 20 ft. deep or less because that's the limit of most drilling rigs used to bore the finished holes. The engineer will take the soil samples back to the lab and analyze them to determine their bearing and expansion characteristics.

These tests, taken together, give a picture of the soil types and conditions at various depths, and of the natural water courses. Armed with this data, the soil engineer makes recommendations about the number and the diameter of the piers needed, and their depth. He specifies the steel-reinforcing requirements and the minimum distance between the piers. In northern California, the typical soil survey costs between $1,000 and $1,500.

Site preparation—The first order of business is to get rid of unwanted plants, loose topsoil, logs and other debris that isn't considered a permanent part of the landscape. Once the lot is cleared, mark stumps, rocks and areas of soft, wet soil with stakes and flagging. The safety of your drill operator could depend on his knowing what's where.

Upslope lots are usually more difficult to deal with than downslope lots. Here's the rule of thumb around here. A foundation on a downslope lot will probably cost twice as much as a foundation on a level site; a foundation on an upslope lot will be three times as expensive. In addition to moving the weight of the entire house uphill, the builder has to deal with the problem of hauling off tons of dirt.

Most soils are compacted in nature, and when drilled into or excavated, they can expand to two or three times their undisturbed volume. A 16-in. dia. by 18-ft. deep pier hole will contain about 1 cu. yd. of compacted soil, which will become as many as 3 cu. yd. of loose soil. So if 15 pier holes are drilled, you might have 45 cu. yd. of loose soil covering the site, or a volume 20 ft. by 30 ft. by 2 ft. deep. This soil can change the contour lines in the site plan, the elevation of the house, the loading on the foundation, and the drainage of the hill. So you have to move it out of there.

In the Bay area, some building departments allow loose soil to be evenly distributed up to 3 ft. deep if the slope is less than 18°. Such soil-dispersal zones must be away from natural drainage channels. On steeper slopes, the loose soil has to be removed or buttressed with a retaining wall to keep it from creeping downhill. Other building departments ask the soil engineer to recommend soil removal or dispersal, then the county inspectors check the project to see if they agree. Each jurisdiction has its own policies governing drilling spoils. You should find out what they are to avoid trouble later.

Most builders stockpile the soil during construction, and then use it for landscaping once the building is finished. If the soil is unusable, or there's just too much of it, it has to be hauled off. We usually rent a dump truck and a tractor with a front-end loader to move the stuff, and take it to the nearest fill site.

Layout—The standard batter board used to locate the corners of perimeter foundations isn't the best solution for layout on slopes. Batter boards don't work well because most pier holes are drilled by a tractor-mounted auger, and as the operator maneuvers the rig around the site, chances are good that the batter boards and some of the survey stakes will be crunched into little splinters. In addition, 12-in. survey pins marking the exact location of piers can shift around in loose topsoils. A D6 Cat weighing around 10 tons may push loose topsoil downhill 1 ft. to 3 ft. as it lumbers around the site. So although the pins may appear to be in the correct relationship to one another, they might have moved downhill with the topsoil. This can cause a lot of problems when you start building your forms.

In order to ensure the correct siting of the house, and to guarantee exact layout points for drilling each pier, I arrange to get my surveyor on the site as the holes are being drilled. The surveyor centers the transit over a survey hub at a fixed elevation away from the drilling area, and then uses a chain (surveyor's tape measure) to locate the corners of the foundation and piers as calculated degree angles are turned. This ensures that no matter what soil movement has affected the preliminary stake positions, each pier hole is located precisely, right at the time it's drilled. The extra cost involved in having the surveyor on site is usually negligible compared to the additional labor and material necessary to do the whole job over again if your pier holes are drilled in the wrong places. If you can't get your surveyor on the site during drilling, the best alternative is to place reference points outside the drilling area to double-check stake location as the pier holes are being bored.

Drill rigs—There are two basic types of drill rigs: tractor or crawler mounted, and truck-mounted. Within these two categories you'll

Dirt flies as a tractor-mounted auger pulls soil from a 20-ft. deep pier hole. Rigs like this can work on slopes as steep as 45° or a little more. The ever-present helper with a shovel positions the auger, plumbs the drill shaft and directs dirt away from the hole.

Rebar for grade-beam and pier cages is assembled on site with jigs nailed to the top of sawhorses. The slots in the jigs position the #4 steel as #3 rebar stirrups are tied in place. Finished grade-beam cages await installation atop the woodpile. A rebar offcut used as a stake holds the woodpile in place.

find rigs that are as different as the people who operate them. The truck-mounted rigs have an auger attached to a boom that can reach several drill targets from one setup point. Such trucks do less damage to a site than a tractor, but they can't negotiate slopes greater than about 30°.

Tractor-mounted augers can drill in terrain that is inaccessible to trucks, and they can work on slopes up to 45° or slightly more. Sometimes an operator will anchor his tractor's winch cable to a stout tree or to a deadman driven into the ground uphill. Some operators carry a power pole on their cat. They auger a hole at the top of the slope for the pole and hook the cable to it. This lets them hang on the side of a steep slope as the holes are drilled. It's a chilling sight.

Drill operators generally charge by the hour, by the footage (the cumulative depths of the holes they drill) or by the job. They base their fees on the capabilities of their equipment, on the difficulty of the job and on the kind of soil they're working in. The rig that's pictured in this article cost $130 per hour, and it took 12 hours to drill 16 pier holes. And with a crawler-type rig, there is an additional delivery charge (often called travel time) tacked onto the hourly rate—$150 for this job. In rocky locations, you can expect a surcharge that will cover the cost of broken equipment, such as auger teeth.

Although these rigs can drill holes up to 20 ft. deep, soil conditions can change every few feet. One hole can be 6 ft. deep and hit rock, and 8 ft. away the next hole might be 18 ft. deep before the same layer of rock is engaged. The variables are surprising, so un-

less you're good at poker (or bridge), do this type of bidding liberally.

During drilling, soil lifted out by the auger is deflected away from the hole by a helper using a shovel (photo, p. 62). The helper will also need a hose to water down the auger bit. Water lubricates the cutting action, reducing the time needed to bore the holes, and lengthens the life of the equipment.

The loose dirt is left at the bottom of the hole because it's too far down to clean it out. Because pier and grade-beam foundations work by friction rather than direct bearing, this loose soil isn't usually a problem. But several years of drought can cause the soil around the piers to shrink and reduce the friction that holds the house on the hill. Enough shrinkage, and the house begins to settle.

You can limit the possibility of soil shrinkage by pouring a few gallons of water and a third to a half-bag of portland cement into each hole after the final depth has been reached and the auger has been removed. Then have the operator re-insert the auger for a minute's worth of mixing. The auger works like a giant milkshake machine to blend cement, water and tailings. When it's lifted out again, the mix should be thin enough to ooze off the drill bit. Once it hardens, this slurry forms a pad that will take direct bearing, and so lessen the chances for foundation settlement during dry years.

Once the holes are drilled, we cover them with plywood to keep out loose topsoil and debris. If there's even a remote chance that small children might wander onto the site, we stake the plywood to the ground and cover it with loose topsoil to make the area uninter-

esting. We also notify the neighbors that there are deep holes covered with plywood on a lot nearby, and we post the area.

Concrete and steel—A pier and grade-beam foundation has a lot of steel in it, and an equally large volume of concrete—100 cu. yd. is not uncommon. These foundations are a lot like icebergs, with their bulk concealed below the surface. With this much steelwork, the builder who sets up an efficient way to fabricate the beam and pier cages can save money.

On a recent job we assembled #4 rebar cages on sawhorses using #3 rebar stirrups and spirals. We easily bent the #3 steel with a rebar bender, and wired the straight pieces in place with standard foundation tie wire (photo above). Some foundations call for heavier steel, and anything over #4 (½-in. dia.) is difficult to bend. If your specs show a lot of #5s, #6s, #7s, consider having the stirrups and spirals fabricated at a metal shop and then delivered to your site for cage assembly.

The foundation can be monolithic, or it can happen in two pours. This latter way produces a cold joint between piers and grade beams. For this job we poured the piers first, and let the cages run long for ties to the grade beams. We suspended each pier cage in its hole with a rebar offcut placed under a stirrup, as shown in the top left photo on the facing page. After the piers were poured, we bent the ends over and tied them into the beam cages (photo facing page, top right).

Forms—When we build our forms depends on the configuration of the rebar. If the grade-beam steel is a series of interconnected cages, we tie them together and then build the forms around them. This allows for adjustment if any piers are out of alignment. The outside of the forms defines the shape of the house, and registers the placement of the mudsills.

If unconnected individual bars are used for the grade-beam steel, it's easier to build the forms first and then place the rebar. Since the forms are built according to the contours of the site, nothing has to be level. We pre-drill the mudsills, insert the foundation bolts and then place them in the wet concrete (photo bottom right). Leveling the house occurs when the foundation walls, sometimes called pony walls, are built.

Pumped concrete—The standard pumping mix in this area contains ⅜-in. pea gravel in a six-sack mix. This aggregate size allows you to use the lower-cost grout pumps, which have a 2 or 3-in. dia. hose. These pumps are pulled behind a pickup, and can pump the concrete 150 ft. up a 45° slope. The six-sack mix has one bag more portland cement than the usual mix. The richer mix is needed to achieve a higher compression rating than you get with the smaller-size aggregate.

You should have water available for lubricating the pump hose, washing loose soil out of the forms and, if pouring the beams separately from the piers, for removing any dirt that may have accumulated on top of them. For cold-joint connections, an epoxy bonder applied to the top of the piers just before the grade beams are poured will help ensure a good bond between the two pours.

Pumpers usually know about how much water they want in the mix, and they will check the mud in the transit mixer for water content. A stiffer mix can clog a dirty hose, and it makes the pump work harder. Since adding more water will lower the psi rating of the concrete, try to hire a pumper who keeps his equipment in good condition. Tell the batch-plant dispatcher you are using a pumper on a steep slope. Adding a water-reducing agent will keep the mix soupy enough to pump and still achieve the required psi compression rating when it cures. Around here this admixture costs only about $2 per cu. yd., and it adds about 1½ in. of slump to a load.

But I still prefer a stiffer mix because it won't run downhill as much as a wet batch. We make an initial pass around the forms, filling them half full. With a stiff mix, the concrete is able to carry some of its own weight by the time we pour in the rest.

If you happen to hear that glorious sound of bursting forms and splitting stakes, halt the pumping immediately. If you wait for 30 minutes to an hour, the mix should set up enough to allow you to continue, despite a weak spot in the formwork. In the meantime, fill other areas of the foundation. Even in mild weather, stripping forms the next day won't keep your engineer happy, but it allows you to get your form lumber back in usable shape. □

Michael Spexarth is a general contractor in El Cerrito, Calif.

A pier cage hangs on a rebar beam (above left) as concrete is pumped into the hole. The protruding bars will be bent over to tie into the grade-beam steel. Once the concrete piers have set up, the pier-cage steel is bent over to tie into the grade-beam cages, as shown at top right. Because there are so many splices involved in the steelwork, Spexarth first installs the cages, then constructs the forms around them. Redwood mudsills are cut to length, and placed near the appropriate footing.

As the wet concrete for the grade beam is screeded flush with the top of the form, the sills are embedded in the wet concrete and checked for side-to-side level. The J-bolt protruding from the sill will be worked down into the concrete, and the nut will be tightened after the beam cures.

Deep Foundations

Using driven piles and grade beams to build on soft ground

by Alvin M. Sacks

If I had a hammer. **Wood piles are pounded in place with a pile driver. They're driven until they won't go any deeper (their "refusal depth"), even though the machine keeps hammering.**

Once in a while you dig a foundation footing and find the soil getting softer instead of harder the deeper you go. If by digging a few extra feet you can reach solid bearing, then the easiest and least expensive solution is to place your footings at that depth. But if the soft soil continues for several feet or more, better alternatives exist. One alternative is a steel-reinforced grade beam supported by some form of concrete or wood piles (drawings facing page).

Concrete or wood?—Concrete piles are usually placed into deep holes that have been excavated with a large auger (like the machine that's used to install utility poles). In very soft soils, drilled holes often collapse before being filled and thus need to be lined with a hollow steel pipe called a casing. The casing may be left in place after being filled, or removed during concrete placement (whether it's left or removed will depend on variables such as the character of the soil and the cost of the steel). A modification of this method eliminates the need for a casing. It consists of a special auger that injects concrete into the hole while simultaneously withdrawing from it.

Wood piles are usually driven in place with a pile driver (photo above), a technique that's especially useful in fill soils that don't contain stumps and rocks. Wood piles are driven down to a "refusal" depth. Refusal simply means that the pile won't go any deeper, even though the pile driver keeps on hammering it. The approximate refusal depth can be accurately estimated using formulas that consider, among other factors, the power of the driver and the resistance of the soil. In general, driven piles have twice the bearing capacity of drilled piles.

Not only do their ends bear on underlying soil or rock, but their sides also develop friction with the surrounding soil.

Coping with soft fill—This case involved three adjacent lots in a draw (a small ravine). Two of the lots had been filled with up to 20 ft. of spoil—dirt that had been excavated and removed from another construction site. The spoil had been trucked in about 20 years earlier and dumped without being compacted. While the fill had naturally consolidated over time, it was still too weak to support the very large, two-story homes that were planned for this site.

Although the underlying soil was probably strong enough to support the houses, it was much too far below the surface. A hole deep enough for two full-height basements would have been needed for conventional footings (such information can be gotten from test pits or borings, or by checking a topographic map that dates from before the fill was brought in). Construction feasibility and cost led the builder to choose treated wood piles and to span the distance between them with grade beams that doubled as wall footings.

Durability of wood—When installing wood in contact with the ground, the question of life expectancy arises. In Venice, untreated piles surrounded by seawater have survived 1,000 years. Untreated piles would not have lasted long on this job, however. Wood can last a long time under water because it's exposed to little oxygen. But these piles would be above the water table, and thus more susceptible to rot. The design called for Southern yellow pine piles treated with chromated copper arsenate (CCA). These are warranted for only 40 years, but if they prove to be as durable as creosote-treated piles they should last much longer—perhaps 75 years. The soil could consolidate and compact enough during that period to support the building, though there's no guarantee. It depends on the soil characteristics at the site. This should be taken into account before deciding whether to use wood piles.

Installing the piles—The piles on this job consisted of 25-ft. long poles that tapered

Drawings: Bob Goodfellow

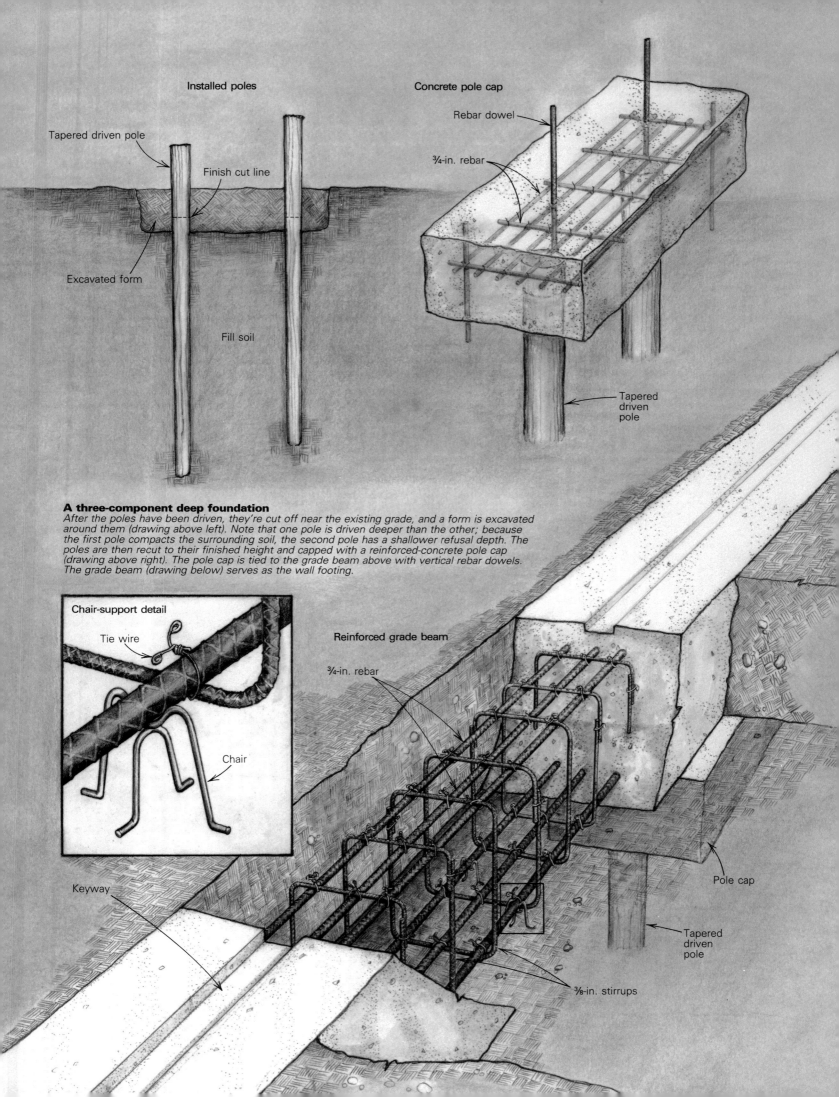

Installed poles

Concrete pole cap

Tapered driven pole

Finish cut line

Rebar dowel

¾-in. rebar

Excavated form

Fill soil

Tapered driven pole

A three-component deep foundation
After the poles have been driven, they're cut off near the existing grade, and a form is excavated around them (drawing above left). Note that one pole is driven deeper than the other; because the first pole compacts the surrounding soil, the second pole has a shallower refusal depth. The poles are then recut to their finished height and capped with a reinforced-concrete pole cap (drawing above right). The pole cap is tied to the grade beam above with vertical rebar dowels. The grade beam (drawing below) serves as the wall footing.

Chair-support detail

Tie wire

Chair

Reinforced grade beam

¾-in. rebar

Keyway

Pole cap

Tapered driven pole

⅜-in. stirrups

Adequate reinforcement. **A grade beam must resist bending forces from any direction. Forming the rebar into a cage provides reinforcement across the beam's entire cross section.**

from 12 inches in diameter at the top to 9 inches in diameter at the base. Because the compressive strength of the wood under axial loads (straight down) is about 1,200 lb. per square inch, each pole could theoretically carry 52 tons. But the large number of poles that were driven under each house meant that each one supported much less than its maximum capacity. Most of the homes' weight was transferred to the soil along the sides of the poles (a phenomenon known as "skin friction"), rather than to the ends.

To provide an adequate margin of safety, two-dozen poles were driven under each house. The structural engineer decided to install the poles in pairs, with 5 ft. between individual piles and 15 ft. between pairs. This close spacing made it easy for the poles to carry the footing, slab, wall and column loads. Pairs of poles were placed under bearing walls, corners and all other concentrated loads.

In pairs such as these, the second pole has a shallower refusal depth than the first, usually by a few feet. This happens because the first pole compacts the surrounding soil as it's driven, which in turn increases the soil's resistance to penetration by the second pole.

After all the poles were driven, the builder used a chainsaw to cut them off just above the existing grade. A backhoe with a narrow bucket then excavated a trench around each pair. Next, the finish elevation of the poles was determined using grade stakes, utility poles and the street as benchmarks. Each pole was then recut to its finish elevation and capped with a concrete pole cap.

Steel-and-concrete caps—A pole cap is a reinforced-concrete pad that's placed over the tops of the poles (concrete is usually referred to as being "poured," but "placed" is the correct term). The cap helps to prevent splitting and curling damage that might result from imposed loads. It also blocks the migration of oxygen from the overlying soil that could abet decay of the pole. Each cap was 2 ft. 8 in. thick and extended a few feet beyond all sides of the poles.

Rebars were laid vertically and horizontally in the pole cap and tied together. The design called for four vertical rebars—two tall ones and two short ones (drawing, previous page). The taller ones would serve as dowels, joining the pole cap to the grade beam that would eventually be poured above it. Such attachment helps to tie all the foundation elements together for resistance to lateral shifting. (Remember that when a foundation wall is backfilled, it acts as a retaining wall as well as a foundation, so it's subject to lateral forces. The greater the backfill height, the greater the lateral force acting on the wall.) The shorter vertical bars acted as grade pegs, marking the finished height of the pole cap.

Forming the grade beams—In this project, the reinforced grade beams also served as wall footings. The engineer called for grade beams at least 12 in. wide by 16 in. to 24 in.

deep. Because they would also serve as a base for 12-in. block, the builder made them about 2 ft. wide. Although form boards were used where the grade changed abruptly, most of the beam was placed in unlined trenches and reinforced with #6, or ¾-in. dia. rebar (rebar is sized by number, as in #6; the number is always a numerator whose denominator is 8). The rebar was held in place by stirrups—⅜-in. dia. rebars that have been bent into squared circles. Tying the long bars to the stirrups with tie wire formed a cage, and guaranteed reinforcement at the top and bottom of the grade beam (bottom drawing, p. 67).

A concrete grade beam, or any other span-ning member, must be designed so that the concrete and steel work together. The steel needs to reinforce the concrete against tension (stretching) and twisting forces. The concrete, in turn, should be thick enough to encase the steel and strong enough to hold it in place. To be protected from corrosion, rebars need at least 3 in. of concrete cover around them on all sides. Rebar should be elevated on chairs—small metal or plastic supports—during concrete placement. Some masons use brickbats to support rebar, but doing so is bad practice because it lowers the beam's tensile strength (see sidebar below). Where two grade beams intersect over pole caps, their rebars

must be tied to each other to maintain the integrity of the reinforcement.

When grade beams double as footings, their surfaces should be roughened to improve adhesion to the foundation wall. For very tall walls or those apt to oppose significant lateral pressures, keyways (2x4s laid in the wet concrete and later removed) or vertically embedded steel dowels make good connecting devices.　□

Alvin M. Sacks has been a builder and consultant for 35 years. He advises on drainage, leakage, paving, roofing, soils, structures and waterproofing in the Washington, D. C., area. Photos by the author.

Brickbats and freefalls

Pole and grade-beam foundations use lots of steel and concrete. Is that much strength needed for a residence? Maybe not, but the reduced quality of much of today's residential construction demands such overdesign. The design sometimes has to compensate for shoddy construction practices.

One common problem is the sloppy way many workers install rebar. To do its job, rebar must be properly located and spaced, adequately tied together, and encased far enough inside the concrete. Brickbats and other odd-sized rough chunks of brick, concrete or stone should not be used as chairs to support rebar (photo below left). Doing so may weaken the concrete.

But the most ubiquitous problem is the excessively wet concrete that's typically used

when placing footings. Concrete develops strength when its components—the aggregate and the paste—are well mixed. Although most contractors want a mix that's wet enough to be poured easily from a ready-mix truck, it's important that the mix not be soupy. The problem with a soupy mix is that the aggregate tends to settle to the bottom, a form of segregation that reduces the concrete's overall strength. This might not be a problem for footings placed on well-consolidated or compacted soil, but it could prove disastrous where the concrete must span a distance, as in a grade beam. A rule of thumb is that if you can pull the concrete with a hoe and have it easily flow around obstacles, it's too loose. For lasting strength, concrete should be stiff enough to require

shoveling around the ditches or inside the forms. Don't use a vibrator to move it around because that can cause segregation. Vibrators are designed only to compact concrete in place by removing entrapped air.

Concrete can also segregate and lose strength when it free-

falls more than about 5 ft. (photo below right). In such cases, using portable chutes or other devices will prevent segregation of the concrete components. If concrete trucks can't be driven all around the site, wood or metal chutes, or concrete pumps equipped with long booms, can be used to deliver the mix.　—*A. S.*

Don't do this. Elevating rebar on brick scraps is bad—but common—practice. Rebar should instead be supported on chairs.

...or this. When concrete free falls more than about 5 ft., the aggregate can settle to the bottom. This seriously weakens the mix.

A Slab-On-Grade Foundation for Cold Climates

Proper drainage (lots of gravel) is the key

by George Nash

The handbooks and structural engineers all say you can't build a slab foundation in frost country and expect it to say put. But about 15 years ago it occurred to me that interstate highways are nothing more than large concrete slabs poured on the ground. They don't have frost walls, yet they don't heave apart. And what about railroad tracks? Why aren't they twisted like spaghetti come spring? Why couldn't the principles of roadbed construction be applied to cold-country slab foundations?

Slab-on-grade foundations are an economical alternative to a full, poured-concrete basement. They require minimal excavation and site disturbance, are quickly laid out, and are easily formed and poured. Also, their suitability as thermal mass in passive-solar and radiant-heating systems is almost unequalled. But

slab-on-grade foundations can be extremely vulnerable to frost damage.

Clay soils (like those of northern New England) are typically saturated with ground water. When this trapped water freezes, the soil expands. If this expansion were uniform, it would pose no appreciable threat to a slab-on-grade foundation. The entire slab would rise and fall evenly, like a boat on the tide. But the perimeter of a slab is subjected to more frost action than the interior. Because concrete is not elastic, any significant difference in movement between two areas will cause a crack. Although frost walls (poured-concrete walls exending to a footing below the frost line) prevent cold from penetrating under the slab, they require so much extra excavation, formwork and concrete that they

offset the savings of building on a slab in the first place.

Frost heave won't occur, however, if the soil can't hold water. Roads and rails are laid upon a base of porous and well-drained material that allows water to move rapidly through and away from it. There's nothing to freeze. With this in mind, I've developed a slab-on-grade system that doesn't need a frost wall for stability. I've continued to refine the system with each new project, but so far there hasn't been any evidence of cracking or movement in any of the half-dozen foundations I've done. Several engineers, including the father of my latest client, have reviewed the system. And while they were skeptical at first, they agree that it will work provided the drainage is good and no water occurs in the soil above the frost line.

Preparing the site—The budget for Walter Breck's house in Fletcher, Vermont, was tight. The site had wet, heavy clay soil underlaid with rock—it would have been hard to find a worse place for a basement. Because of this, and because Breck wanted radiant heating, a slab-on-grade foundation made sense.

We brought in a bulldozer to scrape away the topsoil. The site sloped toward the southeast corner, so the bulldozer actually dug a shallow pit that varied in depth from grade level to about 2½ ft. deep on the north side. Otherwise, we would have brought in clean gravel fill and built up a level pad, compacting the gravel in layers. On a slightly sloping or level site this is not difficult, but in this case, it was easier to dig to level.

The pit was filled to a depth of at least 6 in. with what we call chestnut stone (coarse, 2-in. stone—the same size used for railroad roadbeds). This size stone rakes and shovels hard. Fortunately, the backhoe was already on site digging utility trenches, and we were able to spare ourselves a lot of bone-numbing handwork. I used a hand-held sight level to check the rough grade. Taking the time to do this saves a lot of shoveling later.

We compacted the stone with a gas-powered mechanical tamper (photo facing page) and set up batter boards for the layout strings. Setting all the strings at the same height makes leveling the formboards a simple matter of measuring down from string to board. Here, a transit level is absolutely necessary, particularly for laying out right angles at corners.

A modified grade beam—A thickened edge is recommended for monolithic (one-piece, sans frost wall) slabs. The extra thickness acts as a footing for the load-bearing exterior walls, and the extra depth prevents the foundation from being undermined and provides a surface to install foundation insulation against. Typically, when a slab is poured on a flat compacted base, the perimeter is trenched 8 in. to 12 in. deep and at least a shovel's breadth wide. The slab form boards are set along the outside of the trench, and the inside edge of the trench is sloped upward to the slab depth. This works fine, so long as there is no under-slab insulation and if the depth of the thickened edge does not exceed the width of a 2x12. Also, to minimize rot, wood framing should begin at least 8 in. above grade. But with a 2x12 form board and a 4 in. slab, the finished thickened edge is barely below grade.

I remember one of my early attempts at a slab-on-grade foundation in which I tried to cut and neatly piece together insulation on a crumbly and irregular gravel backslope. During the pour, some of the foam insulation boards tilted, and the concrete flowed under them. Others simply floated away.

On another project I sought to avoid these problems by forming a slab with a 2-ft. turndown, using 2-ft. by 8-ft. strips of plywood for forms. Although I used lots of stakes and braces, I was amazed to discover how much

outward pressure 8 or 9 yards of wet concrete can exert. I didn't lose the wall, but setting the sills to compensate for the free-form curves in the slab edge was rather challenging.

I had read a magazine article about grade beams and it suggested an approach that might resolve the problems of slab edges and insulation. What if, instead of a monolithic pour, I were to form and pour a beefed-up footing and then pour the slab on top of the footing? Reinforcing rods bent at right angles and tied to the wire mesh of the slab would ensure structural connection between the components. Using 2x12s with a 2x4, or even a 2x6, would give a thickened edge of 15 in. or 17 in. and greatly simplify underslab insulation. I tried this for a slab poured over a very gravelly and extremely well-drained site. Gravel was compacted over a base of coarse demolition rubble to ensure drainage. PVC drain tiles were also installed around the perimeter. It worked fine.

Because the base for the Breck foundation was considerably below grade, I decided to double up the 2x12s to form a 23-in. deep beefed-up footing. The formwork was the same as setting footings for an ordinary foundation, except that the beam was only 10 in. wide (drawing below). I set the bottom of the outside form first, leveling it with the batter-board

strings. To accommodate snap ties, I had predrilled holes in the planks 4 in. up from the bottom edge on 3-ft. centers. Next I set up the inside form boards, and marked and drilled corresponding holes. We used round ties designed for a 10-in. wall with 2x4 strongbacks.

Once the snap ties were in place, we tied the first layer of rebar to them. An occasional stake at the center of the sections and at the corners secured the forms while the top section of form boards was added. These we drilled for ties 4 in. down from the top edge.

After the ties were installed in the top section of the form, the forms were tied together and stiffened by strongbacks. The entire form could be moved in or out as a unit to line up with the strings, or raised and lowered to level as needed, using pinchbars and a mattock. Pieces of chestnut stones served as shims. The second layer of rebar was wired to the snap ties, and the 2-ft. long right-angle lengths that would secure the slab mesh were also wired to the cross-ties and the rebar. After making final adjustments, we anchored the forms by backfilling them with chestnut stones and boulders.

Routing mechanicals—The mechanical drawings indicated the exact locations and dimensions for the vent stack, toilet drain, and

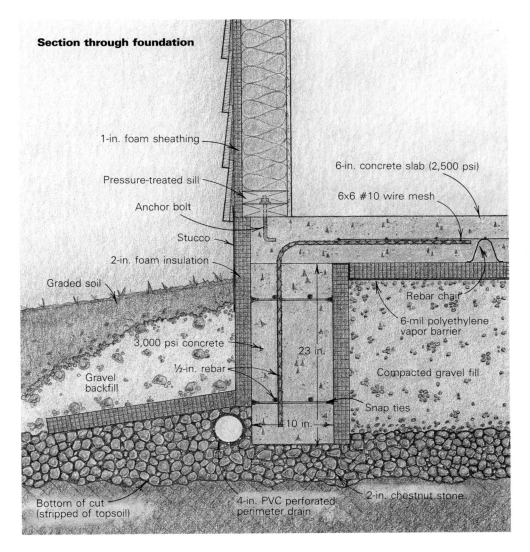

Section through foundation

1-in. foam sheathing

6-in. concrete slab (2,500 psi)

Pressure-treated sill

6x6 #10 wire mesh

Anchor bolt

Stucco

2-in. foam insulation

Graded soil

Rebar chair

6-mil polyethylene vapor barrier

3,000 psi concrete

23 in.

Compacted gravel fill

½-in. rebar

Gravel backfill

Snap ties

10 in.

Bottom of cut (stripped of topsoil)

4-in. PVC perforated perimeter drain

2-in. chestnut stone

other waste and supply lines that would penetrate our beefed-up footing, either horizontally or vertically. Because concrete is rather unforgiving of error, I gave particular attention to these areas, especially when setting the height of the toilet flange relative to the future finish slab (top photo below). I installed a piece of 6-in. PVC pipe as a sleeve at the bottom of the beefed-up footing where the sewer line was to exit. With the exception of the LP gas line, all

utilities entered below the beefed-up footing and could be installed after the beam was poured, and prior to pouring the slab itself.

I used hubless cast-iron for all underslab drainlines. The neoprene fittings allow some flexibility in alignment and are easy to install. For the electric service, I ran 200-amp cable through 2½-in. PVC conduit. It's easier to slip the conduit over the cable one section at a time than to pull or push the cable through

the conduit. The wire is so stiff that 2½-in. conduit barely gives enough leeway. Normally, at least two or three wiring circuits are run under a slab. Using two pieces of 12-3 nonmetallic sheathed cable (often called Romex) in 1½-in. PVC, I can fit four circuits into the space of two. The stouter 12-3 pushes more easily than 12-2 cable. Telephone cables and ½-in. flexible copper LP gas lines fit easily into 1-in. PVC.

I use ¾-in. polybutylene tubing (Qest Systems, Eljer Industries, 901 10th St., P. O. Box 869037, Plano, Tex. 75086) for underslab hot and cold water distribution lines, once again run through 1½-in. PVC. Polybutylene provides security against winter power failures because it can withstand temperatures of -50° F to +180° F without damage. With the exception of the virtually indestructible cast-iron drain lines, all underslab utilities should be replaceable if necessary. We avoided tees and right-angle elbows in either conduits or utility lines because they make it impossible to fish new wires once the slab is poured. Sweep elbows, on the other hand, have gentle curves that won't obstruct cables or pipes, so that's what we used. Teed connections were limited to use above slab.

All measurements were given a final check, and the forms were once again checked against the layout strings before the concrete pour. We used a 3,000-psi mix, poured a bit wet to make dragging easier. The top of the beam was screeded off with a 2x4 and left rough for better bonding to the slab. After screeding, the angled rebars were supported at proper height by stones placed on top of the strongbacks (bottom left photo).

The edges and corners of "green" concrete are easily damaged if forms are removed too soon, so the foundation was left to set up over the weekend before stripping, and then only the inside forms were stripped. The snap ties, now set in concrete, held the outside strongbacks and form boards snugly in place. The outside forms were later built up to serve as forms for the slab.

We laid 2-ft. by 8-ft. panels of 2-in. extruded polystyrene insulation against the inside face of the beefed-up footing, flush with the top edge. The rest of the cast-iron drains and the PVC conduits were then connected to the stubs protruding from the footings, and the sewer line was sealed to its sleeve with Thorobond Waterplug (Thoro System Products, a division of Imperial Chemical Industries, 7800 Northwest 38th St., Miami, Fla. 33166).

I supported the drain line temporarily with wood blocks at a slope of about a 6-in. drop over 10 ft. so that it would exit at the bottom of the beam, roughly 2 ft. below finished grade. Meanwhile, the backhoe dug trenches for water and power below the frost line. The trenches were carefully finished by hand under the beefed-up footing. Then we ran the water line, submersible-pump cable, service cable, telephone and gas lines through their respective conduits. Doing this before the pour makes it easier to troubleshoot any snags.

The vent stack, toilet drain, and other waste and supply lines that penetrated the modified grade beam had to be carefully positioned before the pour.

The top of the modified grade beam was screeded off with a 2x4 and left rough for better bonding to the slab. The strongbacks extend above the 2x12s in order to serve later as supports for the 2x6 slab form. While the concrete cured, the right-angle rebar, which would tie the slab to the beam, was held at the proper angle by stone shims on top of the 2x4 strongbacks.

Clean gravel was compacted in 4-in. lifts to within 2 in. of the top of the beefed-up footing, except for the areas under a bearing post and under the chimney; those areas were left 4 in. deeper to provide extra support. The 24-ft. wide sheet of 6-mil polyethylene used for the underslab vapor barrier required only one overlapping seam at the woodshed/entry ell.

We insulated the slab by laying sheets of 2-in. rigid foam over the gravel and 4 ft. in from the walls. On the south wall, though, they were laid to 8 ft. The ASHRAE (American Society of Heating, Refrigerating and Air-Conditioning Engineers) handbook recommends this technique because the bulk of heat loss occurs at the perimeter of a slab. In retrospect, I probably should have insulated the entire slab, because some heat loss to the earth from the radiant-heating tubes would be likely. Next time.

More pipes, concrete and foam—Because of all the radiant-heat tubing, we decided to pour a 5½-in. slab to guarantee the strength of the floor. The outside strongbacks had been deliberately left projecting about 5 in. above the top edge of the beefed-up footings. Forming for the slab was simply a matter of nailing a 2x6 to them. This also provided another opportunity to adjust for any irregularities in the form work.

Six-by-six wire reinforcing mesh was tied to the angled rebars coming out of the beefed-up footings and supported at proper height above the insulation by rebar chairs. These are U-shaped pieces of rebar that we bought from the local building-supply house. The mesh also held the tubing for the radiant heating system in place (top photo right). Breck's system was designed, and its components were furnished by, Bob Starr of Radiantec (P. O. Box 1111 Lyndonville, Vt. 05851). It consisted of continuous loops of cross-linked polyethylene tubing that distributed hot water from a central manifold to each zone. This system, along with the domestic hot water, was supplied by a 100,000 Btu Polaris high-efficiency (94%), high-recovery LP gas water heater (Mor-Flo Industries, Inc., 18450 S. Miles Rd., Cleveland, Ohio 44128) with an integral heat exchanger (for more on radiant-floor heating, see *FHB* #27, pp. 68-71).

The concrete slab was poured, finished and wet-cured for several days—we sprayed it repeatedly with a garden hose (bottom photo right). After the rest of the forms were stripped, we glued sheets of rigid foam insulation to the outside face of the foundation, flush with the top of the slab. The upper foot or so of the insulation was parged using a foundation coating kit (Retro Technologies Inc., 328 Raemisch Rd., Waunakee, Wis. 53597).

We laid a 4-in. PVC perforated drain pipe around the perimeter at the base of the beefed-up footing, with an outlet to grade at the low corner, and covered it with leftover chestnut stone shoveled up against the bottom of the insulation. This drain is a critical component of the cold-climate slab foundation as it must intercept subsurface and ground water before it can work under the beefed-up footing and up into the gravel slab base.

A horizontal layer of rigid foam insulation was placed over the drainage stone, sloping away from the beefed-up footing, to help protect the bottom of the beefed-up footing and the slab against frost penetration. Finally, the foundation was backfilled with gravel to within a few inches of finish grade, covered with the native soil, and sloped away from the house for positive drainage.

I'm sure that this latest version will be subject to further refinements. But the basic soundness of an insulated slab-on-grade foundation system has been borne out by my experience. □

George Nash is a writer, builder and Christmas-tree entrepreneur in Wolcott, Vermont. All photos by Steve Mandingo.

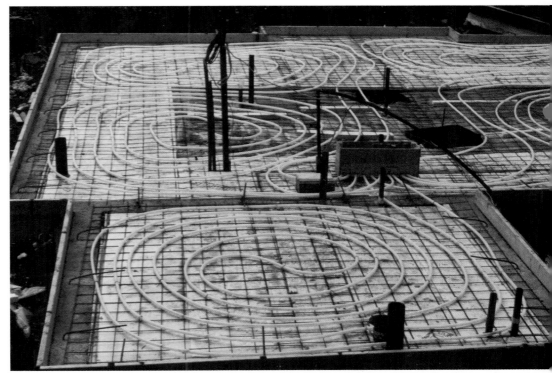

The perimeter of the slab was insulated with 2-in. rigid foam insulation. Six-by-six wire mesh was then laid out and wired to the angled rebar coming out of the beam. Finally, polyethylene tubing for radiant heating was carefully wired in place over the mesh.

In order to keep shrinkage cracking to a minimum, the slab was kept wet for several days after the pour, allowing the concrete to cure evenly.

Foundation Forms that Insulate

Lightweight panels and blocks provide insulation and double as formwork

by Mark Feirer

Houses are a lot tighter than they used to be. Insulation standards have improved, careful weathersealing is the norm, and windows aren't the energy holes they once were. But each time we up the ante on saving energy above grade, energy lost through slab edges and foundation walls becomes a relatively bigger part of the picture. In fact, a foundation can account for as much as 50% of the energy loss in an otherwise energy-efficient house.

So how do you insulate a foundation? There are more options than you might expect, though no one option is a clear favorite of builders. One possibility, however, has considerable promise: the use of foam panels (photo right) or foam blocks (top right photo, facing page) that double as insulation and formwork. The National Association of Home Builders (NAHB) Research Foundation, in fact, considers these systems to be an emerging technology that could significantly change the way foundations are built.

Serious insulation—The insulating systems go by many names. But whether you call them permanent insulated formwork, cast-in-place foundation systems, foam forms or insulating foundation forms (I'll use the latter term), their R-20 to R-56 insulation values go well beyond the puny R-1.6 of an uninsulated 8-in. concrete wall.

The heart of most insulating forms consists either of expanded polystyrene (EPS) or extruded expanded polystyrene (XEPS) insulation formed into lightweight panels or blocks that serve as a form into which concrete can be poured. The forms are typically left in place after the pour, although some can be removed. The outer insulating panel of the LiteForm system, for example, can be stripped prior to backfilling and reused in another foundation, leaving a layer of insulation only on the inside of the foundation. The advantage of doing this will be clear in a moment.

EPS is an open-cell product that rates from R-3.6 to R-4.2 per inch, depending on density (top left photo, facing page). XEPS is a closed-cell product rating about R-5 per inch (middle left photo, p. 75). It's stronger and more water resistant than EPS. Insulating sheets can be made of either material, but one-piece insulating blocks can only be made of EPS because the

Insulating panels. Amid a thicket of braces, workers pump concrete into insulating foundation forms. The system pictured above (R-Forms) is of the panel type: 4x8 sheets of insulation are joined by plastic ties. The exposed ends of the ties (visible here as black squares) provide a base for the attachment of wall finishes. Insulating forms are lightweight alternatives to standard plywood or aluminum forms and offer insulating values over R-20.

complicated shape of one-piece blocks cannot be extruded.

Insulating foundation forms have a 30-year history in Europe, but they've only been on the American scene for a decade or so. Right now, the industry consists of several national companies and a host of regional ones.

The products are most likely to be used by builders who do their own formwork or by owner/builders looking for alternatives to a costly investment in foundation formwork. Foundation subcontractors likely won't have much interest in installing the systems—they already have a big investment in wood or aluminum forms. They might sell insulating forms to supplement their business, though.

Considering the advantages—For homeowners, the chief advantage of insulating form systems comes with the provision of comfortable living areas below grade. Along with high insulating values, thermal breaks are eliminated because the foundation walls are encased in insulation. Condensation below grade is virtually eliminated because the outer layer of insulation isolates the concrete from the relatively cold soil.

As for builders, there are some practical advantages to insulating form systems. Chuck Silver builds houses in New Paltz, New York, and frequently encounters bedrock during excavation. He says an insulated form is easily carved to fit

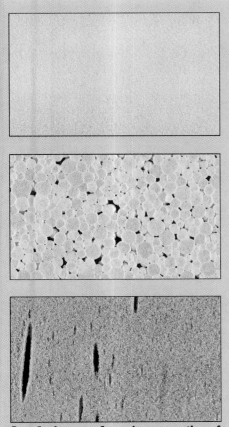

Insulation up close. **A cross section of XEPS (top) and EPS (middle) shows the structural difference between them. Polyisocyanurate (bottom) is used infrequently for insulating foundation forms. Scanned images by The Taunton Press, Inc.**

around rock that can't be removed. Bill Eich builds in Iowa. Part of his marketing strategy is to guarantee a dry foundation, and he thinks that insulating forms help because the outer layer of insulation makes it more difficult for water to reach the foundation.

Builders in northern climates know that cold weather, particularly temperatures below about 40° F, can prevent concrete from curing properly, which in turn leads to a considerable loss of foundation strength. Insulating forms isolate the concrete from the cold and contain heat created during hydration. This means that you can pour in temperatures lower than you'd normally dare to, without precautions such as tenting the forms or including additives in the concrete mix.

Depending on the thickness of the forms, some manufacturers say you can pour in weather as cold as 0° F. I got to wondering how useful this might be, so I called Ken Lewis, a foundation contractor and former builder in Boston, Massachusetts. He investigated lots of insulating form systems: "I'm convinced you really can pour down to zero [degrees], though why anybody would want to is another question. I don't know many people who voluntarily dig foundations in the frost—that's usually emergency work because it's hard on equipment and hard on people."

Another advantage of the forms is that they're very easy to work with, and for owner/builders, this is a real plus. The forms are easily cut to size (a few strokes with a handsaw will do it) and carved out for penetrations (grab the keyhole saw). As with standard concrete formwork, allowance for plumbing and wiring penetrations is created by cutting a hole in both sides of the formwork and inserting a pipe sleeve that will channel concrete around the openings. Most of the insulated foundation systems are relatively easy to manhandle around a site because they're so light—one person can easily lift into place a fully assembled, 4-ft. by 9-ft. section of formwork. Insulating forms built with individual blocks are even easier to work with because they typically include an interlocking edge that makes it virtually impossible to screw up the lay up.

On the downside—For all their advantages, insulating forms aren't perfect. You don't strip the forms, so you'll never know what went on inside. That's why it's a good idea to use a vibrator to consolidate concrete during the pour.

You'll have to protect the portion of the formwork that's above grade because polystyrene will degrade if continuously exposed to sunlight. Eich uses a product called Retro Flex (Retro Technologies, Inc., 328 Raemisch Road, Waunakee, Wis. 53567; 608-849-9000) to protect the exposed surfaces of his insulating foundations. It's a fibermesh product that's troweled over a reinforcing screen. Eich uses two coats of Retro Flex and then paints the stuff. A stuccolike product that can also be used to protect exposed surfaces is Brush-On-Foundation Coating (Dow Chemical Co., P. O. Box 1206, Midland, Mich. 48677; 800-441-4369). And some builders protect exposed polystyrene with pressure-treated plywood.

Polystyrene dings up easily, so you'll have to be a little more careful with these forms than you would with plywood. It's easy to swing a length of 2x around on your shoulder and accidentally puncture a form. Sheets are more vulnerable than blocks to this kind of damage.

But there's one concern that I heard more often than any other while talking to people about insulating forms. It's a problem small enough to crawl through a ⅟₃₂-in. gap.

A problem with bugs?—Termites and other wood-destroying insects are a considerable problem in many parts of the United States, but you wouldn't know this by reading the product literature for insulating forms. Most manufacturers dismiss potential insect damage. They say "polystyrene has no food value for insects" or "no nutritional value." Strictly speaking, they're right; bugs don't go for polystyrene as a food source. They can, however, tunnel through it en route to tastier fare, such as wall studs or the bottom course of your siding. Compounding the problem is that some insulating forms create a concealed path straight to the mudsill. You probably won't spot an infestation until the damage is well along.

Some termite-treatment applicators won't guarantee their work on a house that has any sort of foam foundation insulation, retrofit or otherwise. Silver finally abandoned his use of insulating forms for this reason and now prefers to build a separate, insulated stud wall on the inside of

Insulating blocks. **Reddi-Form blocks (photo above) represent one type of insulating form. EPS blocks with interlocking edges are simply stacked to create formwork.**

Insulating panels. **Most of the insulation for panel systems comes from three manufacturers, so it's the panel tie that distinguishes each panel system from the others. This tie (photo above) holds Lite-Forms together.**

the foundation. He loses some floor space but figures an R-10 stud wall compares favorably in terms of cost and performance.

Some panel systems allow either or both of the insulating surfaces to be removed after the pour. You could, for example, strip the exterior foam surface, and you may even be able to reuse it somewhere else. Terminix, a major U. S. termite-treatment applicator, recommends that any rigid insulation stop below grade. This won't stop insects from getting behind the remaining insulation, but it will allow you to see them when they head for the house.

Getting down to costs—Whatever the advantages of insulating forms, the products are probably going to be judged as much by their cost as by their utility. In April, at the NAHB test house site, I found myself peering into an excavation alongside Bob Syvertson. As the vice president of a long-established New Jersey formwork company, he wasn't convinced by all the talk of R-values and convenience. "I've spent half my life giving prices to builders," he said, "and I know what it's like. Builders are going to look at costs first." And that's where the going gets sticky. It's difficult to sort out the net cost of an insulating foundation because so many factors come into play.

Shipping seems to play a substantial role in determining the cost of insulating formwork. Polystyrene is lightweight but bulky to ship. That's why many of the block systems are shipped KD (knocked down). R-Form blocks, for example, arrive as bundles of flat EPS sheets accompanied by a shipping container of plastic ties. It's up to you (or your least-expensive laborer) to assemble each form by connecting the sides with ties. Other forms (blocks such as Reddi-Form and PolySteel, for example) arrive preassembled. You save time and labor on the job site, but you may pay more for shipping. To get your unit costs down, you have to order more forms. This works to the advantage of builders who can coordinate more than one foundation at a time.

Another factor to consider is that you don't have to strip the forms—that's time and labor savings again. And as a contractor, you don't actually buy insulating foundation forms; in effect, the homeowner buys your formwork and rolls the cost into a mortgage.

But perhaps the biggest problem in comparing the cost of insulating forms and standard forms comes when you try to figure out exactly what you're getting for your money. As the manufacturers of insulating forms are quick to point out, there's more than just formwork going on here.

The biggest value added by insulating foundation systems is that they transform clammy, cold basement walls into a substrate ready for interior finishing in one step. You'd have to insulate the interior walls separately if you poured concrete into standard forms, so add that to your cost comparison. Insulating forms employing metal or plastic ties to hold the sides together are the easiest to finish; you simply fasten paneling or drywall directly to the walls by running self-tapping screws through the exposed end of each tie. Some of the block forming systems don't use ties, however. That's good news when it comes to

assembling the foundation but makes it tougher to secure interior finishes. In those cases, you can either cast in anchoring bolts during the pour and attach nailing cleats to them later, or you can use an adhesive to secure finished surfaces directly to the insulation. Some adhesives will dissolve polystyrene; be sure to follow the manufacturer's recommendations in choosing the right stuff.

Installation—The setup for insulating foundation forms varies according to the system you use. All of the systems (except Forms Plus) are supported on poured concrete footings. Take extra care to ensure that the footings are level because insulating forms don't have much tolerance for anything less. Each form is manufactured to precise tolerances, so flaws in the footing will be transmitted right to the top of the wall.

The bottoms of the forms can be secured to the footing in various ways. R-Form blocks, for example, are placed over short lengths of U-shaped metal tracks that are nailed to the footing; the legs of the U extend upward to prevent the forms from moving. Several other systems fit between a pair of 2x4 plates secured temporarily to the footing with concrete nails.

Block systems lay up much as you'd expect: Head joints are staggered, and courses are interlocked at the corners. The ends and edges of each block are molded to form a T&G or finger joint, which locks adjacent blocks together and prevents concrete from oozing out. The joints have other purposes, too. They make it just about impossible to lay up the blocks improperly, and they stiffen the assembled wall so that only minimal bracing is required.

Vertical rebar in block systems is installed as the footings are poured, then the blocks are stacked over the protruding steel. Horizontal rebar is installed as each course is installed. Where a course ends at a window opening, a foam closing panel is slipped into the exposed end of the block to keep concrete where it belongs.

Panel systems go up much like standard formwork. The difference is that you can assemble the panels in opposing pairs and lift whole sections into place, rather than assembling inside forms and outside forms separately piece by piece. You can build the panels for an entire foundation in your shop while waiting out bad weather, then truck the completed forms to the job site and put them up in a matter of hours.

The key part of any panel system is the tie (bottom right photo, p. 75). These plastic devices connect both sides of the foundation panels, serve as supports for horizontal rebar and provide anchoring locations for interior and exterior finishes. Each tie ends in a flat plastic plate that's left exposed on the outside surface of the formwork. Wall finishes are simply screwed to the plates with self-tapping screws.

With some panel systems, you purchase standard insulation locally, then drill each sheet to accept the proprietary ties. This offers the benefit of the best local prices and easy availability, and it also lets you choose between EPS or XEPS panels in any thickness and size. On the other hand, you'll spend a bit of time drilling the panels for ties. In the case of R-Forms, for example, each 4x8 panel requires 18, ½-in. holes. But there's a way to make the work go quickly. The R-Form assembly manual shows how you can gang-drill stacks of panels with the aid of a spade bit, a plywood template and a makeshift worktable.

Individual panels can be locked together with plastic splines that slip over the vertical joints. Special inside and outside corner pieces secure intersecting panels to each other. As for bracing the walls, the systems seem to vary considerably in the amount that's required. Some systems are braced by a maze of wood walers and cleats (photo right), and others need only a bit of stiffening at the top edges and occasional braces in the field of the wall (middle photo, p. 78). Several manufacturers provide an aluminum bracing system that eliminates the need for wood bracing. If you plan to install more than a few

Source list

Insulating wall forms come from a variety of sources; some can be purchased directly from the manufacturer; others must come from local or regional distributors. The following list includes all the manufacturers I've been able to locate, as well as other sources of information about insulating forms. —M. F.

American Polysteel Forms
3-10 Insulated Forms (distributor)
P. O. Box 46790, Omaha, Neb. 68046
(800) 468-6344
EPS blocks.

Amoco Foam Products Co.
400 Northridge Road, Suite 1000,
Atlanta, Ga. 30350-3323
(404) 587-0535
(800) 241-4402
Manufacturer of XEPS panels.

Branch River Foam Plastics
15 Thurber Blvd., Smithfield, R. I. 02917
(401) 232-0270
EPS panels.

Cubic Industries, Inc.
4931 Meinders Road, McFarland, Wis. 53558
(608) 838-6607
EPS blocks.

Dow Chemical Co.
P. O. Box 1206, Midland, Mich. 48677
(800) 441-4369
Manufacturer of EPS sheets.

EnerComp, Inc.
68 Pine St., Natick, Mass. 01760
(508) 655-2565
EPS blocks.

Energ-G Block Corp.
2850 S. Roosevelt, Suite 102,
Tempe, Ariz. 85282
(602) 470-0223
Blocks made from closed-cell polyurethane.

Faswal
Box 189, Windsor, S. C. 29856
(803) 642-9346
Insulating blocks made from a wood chip/cement mixture.

Forms Plus
P. O. Box 3166, Yuba City, Calif. 95922
(916) 671-4570
XEPS panels.

K & B Associates
20 North Railroad, San Mateo, Calif. 94401
(408) 279-1222
EPS blocks.

Lite-Form, Inc.
1210 Steuben St., P. O. Box 774, Sioux City, Iowa 51102
(712) 252-3704
EPS and XEPS panels.

R-Forms, Inc.
10800 N. Military Trail, Suite 216,
Palm Beach Gardens, Fla. 33410
(407) 624-2515
(407) 624-3247 FAX
EPS blocks.

RASTRA Building System
6421 Box Springs Blvd., Riverside,
Calif. 92507
(714) 653-3346
EPS/cement blocks.

Reddi-Form, Inc.
250 Canal Road, Fairless Hills, Pa. 19030
(215) 295-8884
EPS blocks.

Therma Manufacturing
649 University Ave., Los Gatos, Calif. 95031
(408) 395-8183
EPS blocks.

UC Industries, Inc.
3 Century Dr., Parsippany, N. J. 07054
(201) 267-1605, x4376
Manufacturer of XEPS panels.

Bracing for a pour. The amount of bracing required with insulating forms varies considerably. The formwork shown above requires numerous vertical cleats and horizontal walers, as well as kickers run to ground stakes. Contrast this to the formwork shown on the following page (center photo).

Lite-Forms. Minimal bracing is needed for this system due to the great number of ties (top photo) used to hold the sides together. As with most insulating form systems, the concrete is generally placed by pump (photo above). Pumping concrete offers greater control of the pour and minimizes damage to the forms.

Forms Plus. This system is intended primarily as formwork for slabs with monolithic footings. A PVC sheet protects the XEPS insulation and forms a termite shield at the top of the form.

insulated foundations, the extra cost of such a bracing system may be worthwhile.

Concrete mix and placement—When you slow down concrete's rate of cure, the result is stronger concrete. Insulating forms slow the curing. Some of the systems, particularly those with a complex network of inner channels, should be poured with low-slump concrete (slump is a measure of the consistency of concrete; low-slump concrete flows easier than high-slump concrete does). Adding water is one way to lower the slump, but this weakens the concrete. Also, the additional water adds weight to the mix, which in turn increases pressure within the forms. If there's too much pressure, the concrete could blow out part of an insulating form. The best way to affect the slump of concrete is to include chemical additives in the mix. These additives, available from the concrete supplier at extra cost and typically added at the plant, change the slump without affecting strength.

Many manufacturers of insulating forms recommend pumping concrete into the forms or using some sort of conveyor system, rather than chuting concrete directly from the ready-mix truck. Pumping or conveying the concrete offers more control of the pour and avoids damage to the top surface of the formwork. A 2-in. or 2½-in. dia. hose generally works best; 3-in. and 4-in. hoses can overfeed the cavities.

Depending on the height of the foundation wall, you may have to pour the concrete in lifts (layers). At some height, any form, including wood forms, won't be able to withstand the pressure of all the concrete inside. Some of the insulating form manufacturers recommend lifts of 4 ft. or so while some forms are warranted to withstand at least 10 ft. If you do have to pour the wall in two lifts, the first one needn't cure completely before the second one is poured, but you'll have to check with an engineer or a concrete specialist to determine just how long you should wait between lifts.

Sampling the systems—A number of companies manufacture insulating foundation forms (see sidebar, previous page). Rather than describe every system, however, I'll sketch six systems that typify the range of products now on the market.

Lite-Form. The Lite-Form system consists of 4 ft. by 8 ft. sheets of EPS held together by anchor ties made from reprocessed polypropylene. One advantage of the Lite-Form system over other panel systems is that you can use any EPS or XEPS insulation; all you buy from Lite-Form is their expertise and their form ties, so you can shop local sources for the best price on insulation. Designers of the NAHB Resource Conservation Test House, for example, specified a Dow XEPS panel that contains 10% post-consumer recycled polystyrene (middle photo, left). The foundation cavity—the space between the forms— ranges in thickness from 4 in. to 16 in. The ties are spaced 8-in. o. c. throughout the wall and extend through the insulation on both sides of the wall. Form bracing is minimal (top photo, left): 2x4 kickers hold the formwork steady, and a simple collar of 2x4s stiffens the top edge of the forms. You can also use the company's proprietary bracing system.

R-Forms. Unlike the Lite-Form panels, R-Forms are fabricated at the factory. The XEPS panels used most often with the R-Form system result in an R-20 wall, but insulating values ranging from R-8 to R-56 are possible. The 4 ft. by 8 ft. panels are assembled on site with plastic ties (photo p. 77). Depending on the design of the wall, ties can be placed 12-in. o. c. or 16-in. o. c. vertically and horizontally, and the foundation cavity can range from 4 in. to 10 in. thick. The company's bracing system includes tubular aluminum kickers and vertical stiffeners, but you can also use 2x4 braces.

Forms Plus. Forms Plus is the only system I know of that was designed as stay-in-place formwork for slab-on-grade foundations with monolithic footings (bottom photo, facing page). It's a new product that should be on the market by the time you read this. A high-density XEPS sheet forms the outside face of the formwork while the excavation itself forms the inside face. The outer surface of the insulation is protected by an exterior-grade PVC sheet. The PVC wraps over the top edge of the insulation, then partly down the backside to form an integral termite shield that keys into the concrete. Even if bugs do get behind the insulation, they'll find it tough to get past the shield.

The PVC sheet is adhesive-bonded to the insulation at the factory. As a shield, it protects above grade portions of the insulation from weather, errant stabs of pruning shears and other threats. As a disguise, the PVC has a matte gray finish that makes it look like concrete to the casual observer; it can also be painted with water-based paint or covered with stucco. The PVC also gives the forms additional rigidity. This, along with careful bracing, helps the forms withstand the pressure of the wet slab.

After excavating for the foundation walls, the builder slips forms into the trench and secures them with wooden stakes and kickers. The top edge is carefully leveled, and seams between forms are sealed with PVC cement. A serpentine dovetail key in the back of the product provides a mechanical connection between the form and the concrete. Why serpentine? Straight grooves would weaken the form just as saw kerfs weaken a sheet of plywood.

PolySteel. The PolySteel system consists of 48-in. by 16-in. EPS panels joined into blocks by proprietary ties, resulting in an R-22 wall. What sets

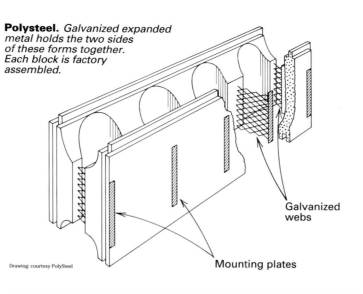

Polysteel. *Galvanized expanded metal holds the two sides of these forms together. Each block is factory assembled.*

Galvanized webs

Mounting plates

Drawing: courtesy PolySteel

RASTRA. **These block forms consist of PVC particles mixed with cement and water. They can be handsawed and shaped with rasps.**

Reddi-Form. **The horizontal passageways within these block forms are evident here. The interlocking blocks require little bracing.**

these blocks apart from others is that the ties are galvanized metal lath rather than plastic, and they're integral to the form instead of installed on site (drawing left). Five of these ties run vertically through each block and terminate in 2-in. by 16-in. galvanized metal plates that provide a base for attaching interior and exterior wall finishes. The inner surface of each form is molded to create a network of horizontal and vertical columns running throughout the wall. The columns are connected by concrete webs, resulting in a monolithic concrete wall that uses 25% less concrete than a standard wall of similar thickness. The wall still provides enough strength to support a typical house.

RASTRA North America. If you associate EPS only with coolers and hot cups, the folks at RASTRA will change your mind in a hurry. Their block product is EPS alright, but that's not the whole story. EPS particles are mixed with cement and water to form a block that can withstand hammer blows and bugs (photo left). Like the other insulating form blocks, the RASTRA block is simply stacked into a wall and concrete is poured into the cavities. The complete wall weighs in at R-44. Unlike any of the other insulating foundation systems, however, a poured RASTRA wall can pass the Underwriter's Laboratories' two-hour fire test without any supplemental surface treatment. The blocks can be cut or shaped with standard hand tools.

Reddi-Form. When assembled into a wall, these R-21 EPS blocks create horizontal and vertical passageways (bottom photo, left). The resulting wall is not monolithic, however. If you removed all the EPS from a poured foundation, you'd see a grid of vertical concrete columns $9\frac{5}{8}$ in. apart, connected by horizontal concrete beams 12 in. apart. The foam webs that connect each face of a block run clear through the product.

The columns and the beams have diameters of 6 in., so you have to be careful in placing steel reinforcement. The company provides formed wire "bridges" that provide support for the rebar. But an engineer must specify the exact placement and type of rebar because the Reddi-Form system is not pre-engineered. The forms are warranted for lifts of up to 9 ft. in height. □

Mark Feirer is the former editor of Fine Homebuilding. *Photos courtesy of manufacturer except where noted. For more on the subject of foundation insulation, see* The Builder's Foundation Handbook *by John Carmody, Jeffrey Christian and Kenneth Labs (reviewed in* FHB #74, p. 98).

Permanent Wood Foundations
A cost-effective and energy-efficient alternative to concrete

by Bill Eich

The permanent wood foundation (PWF) is an engineered foundation system that's made with perservative-treated wood, and my company has been installing them for more than ten years. Though we were initially skeptical, our experiences have made us believers in the system. As a company that specializes in the design and construction of energy-efficient, airtight housing, we require basements that are warm, dry and structurally sound. The PWF has all of these qualities, and at a cost that's competitive in this area with poured concrete and concrete block.

PWFs have been available for many years (they were originally known as All Weather Wood Foundations; see *FHB* #5, pp. 40-42), yet they account for a mere 5% of the U. S. market and about 20% of the Canadian market. Buyer resistance isn't the problem, though; When both systems are explained and the benefits compared, 75% of our clients in the $100,000 to $200,000 price range choose wood over concrete. The main obstacles to more widespread use of the system have been builder resistance and a powerful concrete-industry lobby that has slowed code approval in many jurisdictions. In spite of these obstacles, however, PWFs are now approved by all major codes in America and Canada, though a few cities still prohibit or limit their use. In this article I'll describe why I think the PWF is such a good system. I'll also give an overview of how to install one, including some tricks we've learned over the past decade that aren't included in the official installation guides.

Low cost, high comfort—From the consumer's viewpoint, the benefits of a wood foundation are warmth, dryness and finishability. A wood-foundation basement can be easily converted to an inexpensive living space that's as comfortable as the main level. In areas where residential construction costs average $60 to $70 per sq. ft., a wood basement can normally

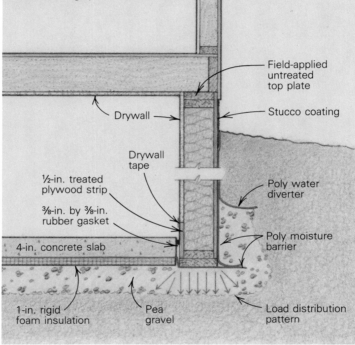

A wood foundation is basically a below-grade treated-wood studwall. The differences between this system and an above-grade wall consist of details that are meant to divert ground water and resist backfill pressures. Proper detailing will ensure a comfortable, low-humidity basement that's easily converted to finished living space.

- Field-applied untreated top plate
- Stucco coating
- Drywall
- Drywall tape
- ½-in. treated plywood strip
- ⅜-in. by ⅜-in. rubber gasket
- 4-in. concrete slab
- Poly water diverter
- Poly moisture barrier
- 1-in. rigid foam insulation
- Pea gravel
- Load distribution pattern

be converted to finished space for $10 to $20 per sq. ft. Even in an unfinished basement, R-19 insulation batts installed from floor to ceiling will reduce both heat loss and utility bills. And the superior drainage features of the PWF ensure a dry basement storage area that's free from mold and mildew.

From a builder's perspective, the benefits of a PWF are cost and control. It costs us $20 to $25 per lineal foot to install a wood foundation, while our subcontractors charge us $30 to $35 for a masonry wall. Using our own framing crew to install foundations gives us much more scheduling flexibility than if we had to depend on a masonry subcontractor. And a PWF can be installed during the winter without expensive shelters or temporary heating units. At most, a little preplanning is required. Spreading a layer of straw or foam insulation over the building site keeps the ground from freezing too deeply for excavation. This lets us start jobs with wood foundations in

January or February that we would otherwise postpone until April.

Two of the first five PWFs in America were built and installed in 1971 here in Spirit Lake, Iowa, by Citation Homes, one of the nation's leading suppliers of factory-fabricated wood-foundation panels. Those foundations are performing today, more than 20 years later, as efficiently as the day they were installed. And though no one knows what the ultimate lifespan of a typical foundation will be, at least one company guarantees them for 75 years.

Sticks and stones—A PWF is a load-bearing, lumber-framed wall that's sheathed with plywood. It sits on a concrete footing or a bed of gravel or crushed stone (drawing left). A wood foundation can be framed with 2x6s or 2x8s, depending on the particular loading requirements. It goes together much like a standard studwall, with studs, plates and plywood. But there are important differences. The need to resist backfill pressure creates more critical loading and stress requirements than are present above grade. Because of this, it's important to use the proper fastening and blocking techniques. In fact, most problems with wood foundations can be traced to improper installation by inexperienced workers.

Wood-foundation problems are less expensive to correct than those involving masonry foundations. Even so, builders just learning the system should start by using prefabricated foundation panels that have been properly engineered at a component manufacturing facility. Only after a great deal of experience with factory-fabricated foundation panels should one consider site-building a PWF—it's too easy to miss a small detail and have it come back to haunt you. Though there's no central source of information about panel manufacturers, two good companies I'm aware of are Citation Homes (1100 Lake St., Spirit Lake, Iowa 51360; 712-336-2156) and Permanent Wood Foundations, Inc. (P. O. Box 819, Flint, Mich. 48501; 313-232-5099). The former will ship panels

Drawings: Vince Babak

anywhere from Colorado to Ohio, the latter ships nationwide.

The foundation-grade lumber used in a PWF is pressure-treated with a chromated copper arsenate (CCA) solution to a retention level of .60 lb. of chemical per cu. ft. of wood—50% more preservative than the codes require for ground-contact lumber. Although the chemical is water soluble during treating, it permanently fuses to the wood cells as it dries, making it clean and safe to use. Southern yellow pine allows the highest level of preservative penetration, making it the material of choice for most treated wood foundations (for more on preservative-treated wood, see *FHB* #63, pp. 61-65).

Laying the footings—Regardless of whether you use site-built or prefabricated panels, the first step is to dig the basement and footings. The basement is excavated 10 in. to 11 in. below the finished basement-floor level. To minimize backfill pressure on the walls, we try to limit the overdig around the perimeter (the extra width and length needed to facilitate working around the outside of the foundation) to between 12 in. and 18 in., particularly where we have a full 7 feet of backfill height. We dig our foundations with an end loader, which often means excavating a ramp into the hole. We try to place this ramp at the lowest point of the final grade around the foundation to further reduce backfill pressure.

After the excavation is complete, we set the perimeter stringlines. Then 12-in. long, treated 2x2 grade stakes are driven every 4 ft. to 6 ft. around the foundation perimeter, just inside the stringlines. We use a transit to set the tops of these stakes 4½ in. below the finish floor level, which leaves them about 6 inches above the basement subgrade. The entire subgrade is then covered with 5 inches to 7 inches of washed pea gravel, which is spread with a skid loader (such as a Bobcat). The perimeter needs to be very level, so it's floated with a 2x4, using the grade stakes as a leveling guide (top photo). Drain tile isn't required with a wood foundation: the pea-gravel base acts as a huge drain tile. We usually install a sump basin in a utility room; any ground water flows into the basin and is pumped away.

Next, we lay the treated footing plate around the perimeter. For 2x4 foundation walls, a 2x6 footing plate is required, while 2x6 walls need a 2x8 plate. The footing plate is nailed down into the 2x2 grade stakes (middle photo). The grade stakes serve no structural function; they are merely a convenience to level the pea gravel and hold the footing plates during the erection process. We've tried several other techniques but the 2x2s work best.

Many builders get nervous about the prospect of letting all foundation loads bear directly on the pea gravel, with no concrete footing. To understand why the technique works, remember that when confined in an enclosed space—such as a basement ditch—pea gravel is non-compressible. You can prove it to yourself by taking a 5-gal. bucket, filling it with pea gravel and striking off the top. Regardless of

Moisture control is crucial to a properly functioning wood foundation (photo above). First, the subgrade is covered with pea gravel, and the perimeter is leveled using grade stakes as a guide.

A treated footing plate is then installed (photo above) over a 2-ft. wide poly strip and is nailed down into the grade stakes.

This strip is tape-sealed (photo below) to a poly moisture barrier installed over 1 in. of rigid-foam insulation.

how much it's shaken or tamped, it won't settle or compress. The pea gravel spreads the foundation wall load from the footing plate to the subgrade below at a 45° angle. The bearing capacity is thus a function of the depth of the pea gravel: an 8-in. wide footing plate on 6 inches of pea gravel will spread the load over a 20-in. wide area, just like a 20-in. by 8-in. concrete footing.

A 2-ft. wide strip of 4-mil poly is installed beneath the footing plate. It serves as a capillary moisture break. This strip is sealed with housewrap tape to a continuous poly moisture barrier under the entire basement floor (bottom photo, previous page). An ordinary 6-mil poly meets the PWF specification guidelines, but we prefer to use a higher quality 4-mil product called Dura-Tuff (Yunker Industries, 200 Sheridan Springs Rd., Lake Geneva, Wisc. 53147; 414-248-6232), which is both puncture resistant and UV stabilized. If Dura-Tuff is unavailable, we use a product called Rufco (Raven industries, P. O. Box 1007, Sioux Falls, S. D. 57117; 605-336-2750) that has similar properties. We also place 1 in. of rigid foam under all of our basement floors, laying the foam directly over the pea gravel and beneath the moisture barrier. The foam isn't required by the PWF specifications, but using it is good construction practice for comfort, moisture control and energy efficiency. If you don't use foam, you'll have to screed the pea gravel to the top of the footing plate or pour the basement floor 4½ in. thick.

Slab first—At this point, many builders fabricate and erect the foundation wall panels, leaving the basement floor for later. We choose to pour our basement floors first because concrete is a much better work surface than pea gravel. We place our floor forms directly on the footing plate. The edge of the slab is held back ½ in. from where the inside edge of the wall will be. This leaves room for a ½-in. treated plywood strip (bottom photo, right). We used to install this so that the top edge was flush with the concrete floor; now we let it protrude about 2 in. This lets us tape the plywood to the finish drywall, completing the wall air barrier. Sure-Seal, a ³⁄₈-in. by ³⁄₈-in. saturated urethane foam gasket (Denarco Sales, 12710 Idlewild, White Pigeon, Mich. 49099; 616-641-2206) completes the air barrier between the wall and the slab, as well as reducing radon-gas penetration. The concrete floor must extend 1½ in. above the bottom of the studs—it keeps the bottom of the foundation wall and studs in place during the backfill.

Framing the walls—Once the basement floor is finished we begin framing the walls. All fasteners must be corrosion resistant. PWF specifications require double hot-dipped galvanized nails above grade, stainless-steel below. For power nailing, we use type 304 or 326 stainless-steel nails or staples. Senco Products, Inc. (8491 Broadwell Rd., Cincinnati, Ohio 45244; 800-543-4596) makes stainless-steel fasteners that meet the specification for air nailers.

There's a fair amount more crown and twist in treated lumber than in standard framing stock. To compensate for this, we figure on a little more waste and make sure that all crowns face toward the exterior. Doing so sets the studs in opposition to backfill pressure which, in turn, helps straighten the studs. We save the straightest pieces for the corners and plates. Studs are placed with the cut ends at the top of the wall so that we won't have to re-treat the occasional piece that hasn't been fully penetrated by preservative. The wall is sheathed with ½-in. treated plywood, and all joints are sealed with butyl caulk.

Stud size and spacing vary with material grade and backfill depth. In general, though, 42 inches of backfill will require a 2x4 wall framed at 12 in. o. c., 64 inches of backfill a 2x6 wall 16 in. o. c., and 84 inches of backfill a 2x6 wall 12 in. o. c. It's possible to engineer a wood-foundation wall with a full 96 inches of backfill, but the added cost of the required 2x8 framing makes the finished foundation more expensive than concrete.

Window openings require one extra king stud for each stud that's been cut. The rough sill piece on windows must be doubled for window openings up to 6 feet wide and tripled for openings wider than 6 feet. The sill distrib-

Energy detailing. Eich lets a ½-in. treated plywood strip protrude 2 in. above the basement floor (photo above). It is later taped to the drywall. This, along with a rubber gasket, completes the foundation air barrier. Before the wall is nailed in place (top photo), a small hydraulic jack compresses the rubber gasket between the wall and the basement floor slab to ensure a tight seal at the joint.

utes the backfill pressure from the front of the window to the king studs at the sides.

After the walls have been framed and sheathed, we fasten the self-adhesive Sure-Seal to the side of the slab and then stand the walls upright. Here we see another advantage of pouring the slab first. Before nailing the walls to the footing plate, we push them against the slab with a small hydraulic jack (top photo, facing page). This not only straightens the bottom plate but it compresses the rubber gasket, ensuring a tight seal between the wall and the slab. A second, untreated top plate is also nailed on at this point.

The first-floor deck needs to absorb and distribute any backfill loads. Because of this, the foundation can't be backfilled until the floor is complete. Care must be taken to toenail each floor joist to the top plate with three 10d nails. The rim joist is also toenailed at least every 12 in. On walls where the floor joists run parallel to the foundation, 2x10 blocking must be installed—24 in. o. c. maximum in the first joist bay and 48 in. o. c. in the second (photo below right). The blocks are secured to the foundation wall with metal framing anchors. The subfloor is glued and nailed to the blocks as well as to the joists. This is an area where proper installation is critical: a poor job of fastening the floor joists or the blocking will cause the wall to fail.

Special considerations—Some details that require no special thought or care with masonry foundations are quite different with wood. A basement stairway that runs along an exterior foundation wall is a good example. Because the stair opening will prevent the floor deck from absorbing the wall loads at that point, a beam must be built to transfer the horizontal loads around the opening to the floor. We build this beam by gluing and bolting six 2x6s flat on top of the foundation wall along the stairway. The beam extends about 2 feet beyond each end of the stairway (drawing right).

Uneven backfill heights are another potential problem, particularly in homes with an 84-in. backfill height at the front of the house and a walkout basement at the rear. In these cases, interior shear-wall panels spaced 15 feet to 20 feet apart will help the foundation resist any racking forces (top photo, next page). A shear wall is a 4-ft. section of wall perpendicular to the foundation wall that's sheathed with ½-in. plywood or oriented strand board and nailed 4 in. o. c. Shear walls are lag-bolted up into the floor joists and down into the basement floor. We try to place these walls where a future interior wall may be located. An alternate, but more expensive technique, is to bury a concrete deadman outside of the foundation wall, tying it to the studs with re-rod stirrups. Finally, whenever a garage floor, sidewalk or patio is poured next to a wood foundation, tying the slab to the foundation with lag bolts or re-rod stirrups will stiffen the top of the foundation (bottom photo, next page). Whatever the permanent support sys-

tem, it's a good idea to add a few extra temporary braces to the inside of the foundation before backfilling.

After the walls have been raised and braced, the foundation is wrapped with an 8-ft. wide sheet of poly. We prefer Dura-Tuff for this application, for the same reasons we use it under the slab. The poly acts as a water diverter, an area of low friction that will direct ground water to the pea-gravel footing. The first 12 in. around the foundation is backfilled with pea gravel and topped with a poly strip. We backfill with a skid loader and make sure that the walls are braced. The bracing prevents large rocks and clods of soil from bouncing off the walls and causing them to bow inward.

The above-grade portion of the treated plywood can be finished in a variety of ways. Although lxl2 cedar, ¼-in. cement board and fiber reinforced plastic are all commonly used

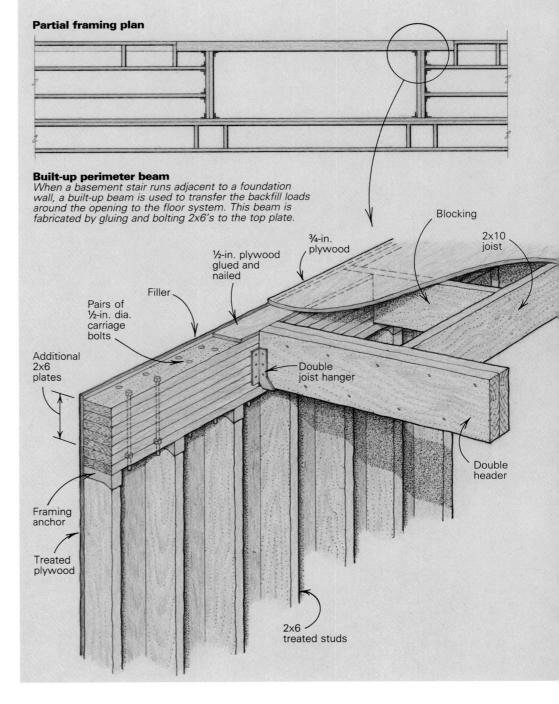

Partial framing plan

Built-up perimeter beam
When a basement stair runs adjacent to a foundation wall, a built-up beam is used to transfer the backfill loads around the opening to the floor system. This beam is fabricated by gluing and bolting 2x6's to the top plate.

¾-in. plywood

Blocking

2x10 joist

½-in. plywood glued and nailed

Filler

Pairs of ½-in. dia. carriage bolts

Double joist hanger

Additional 2x6 plates

Double header

Framing anchor

Treated plywood

2x6 treated studs

To absorb backfill loads, blocking is needed near foundation walls as the underside of this first floor shows. The subflooring is then glue-nailed to the joists and the blocking.

treatments, our first choice is a brush-on stucco coating called Retro Flex (Retro Technologies, Inc., 328 Raemisch Rd., Waunakee, Wisc. 53597; 608-849-9000). It's easily applied and gives the walls the appearance of concrete.

The bowed and twisted pieces of treated wood discarded earlier can be used for bearing-ledge panels or garage foundation panels. They get buried in the ground and don't have to be straight. A bearing-ledge panel is simply an open 2x4 stud panel with studs 16 in. o. c. and a single top and bottom plate (photo above). It's attached to the outside of the basement wall after the poly water diverter is on. Its purpose is to carry brick veneer, a sidewalk, masonry steps or a garage floor.

Garage foundations or slab-on-grade perimeter foundations can also be built with treated wood. These typically consist of 2x4 studs, 36 in. to 48 in. long, with a double top plate and a single 2x6 sill plate that also serves as a footing plate. To provide some shear strength, the top 24 in. of the foundation panels are sheathed with ½-in. plywood. Garage foundations usually protrude 6 in. to 12 in. above the floor, so 12 in. of plywood is also nailed to the inside to allow a cleaner concrete job. For a slab-on-grade system, the concrete floor is poured flush with the top of the wall and the inside plywood is eliminated. Whether garage or slab-on-grade, we brace these walls thoroughly before backfilling—there's no slab to hold the bottom in place and no floor framing to hold the top. Even with the extra brac-

Where backfill heights vary widely, interior shear walls resist backfill pressure (top photo). Tying the foundation to masonry adds even more stability. The box that's being framed outside the foundation is a bearing-ledge panel. It will eventually carry a concrete stoop. Re-rod stirrups (photo above) will secure the foundation to an adjacent driveway or slab.

ing, we must still sometimes re-align the garage panels with the skid loader after they've been backfilled.

Early experiences—A final, rarely mentioned advantage of wood foundations is that they're more forgiving of error than are concrete ones. One foundation we installed was backfilled during the winter with a combination of snow and large dirt clods. When the spring thaw came, a 2-ft. deep trench appeared all around the house, followed immediately by 2 in. of rain. As I approached the house after the rain stopped, I could see the water line where the basement ditch virtually formed a moat. Expecting the worst, I found instead a perfectly dry basement and a hard-working sump pump.

Another wood foundation we installed was quickly capped off with a floor system early one winter in a snow storm. We didn't return to frame the house until spring, when we noticed that one of the end walls had bowed in about 2 ft. at the top. After closer examination, it was obvious that in their haste to get out of the snow, the crew had forgotten to install the metal framing anchors on the endwall blocks. I brought in the backhoe, dug out the backfill, pushed the wall back out with one hand, nailed the framing anchors in place properly and backfilled again. The problem was fixed in one hour at a cost of $45. A failing concrete wall would have been much more expensive to fix. □

Bill Eich is a custom-home builder from Spirit Lake, Iowa, and is president of the Energy-Efficient Building Association. Photos by Charles Wardell. For additional technical information, Permanent Wood Foundation: Guide To Design and Construction, *58 pp. is available from the American Plywood Association (P. O. Box 11700, Tacoma, Washington 98411; 206-565-6600).*

Wood Foundations
Pressure-treated studs and plywood make an economical system for owner-builders

by Irwin L. and Diane L. Post

We first heard about All Weather Wood Foundations (AWWF) in 1978. We later learned more about this pressure preservative-treated system, and when we began to build our own house we chose it over the more common poured concrete and concrete-block foundations. There were several reasons for our choice.

First, the rugged winters here in the mountains of Vermont made the insulation of the house one of our primary concerns. Since we planned our basement to be living space, we wanted it to be as well insulated as the rest of the house (fiberglass to R-26 in the walls and R-38 in the top floor ceiling). AWWF walls are built of studs, and they can be insulated with fiberglass as easily as can any stud wall. With concrete foundations, a thick layer of insulation significantly reduces the usable space inside the basement, and it's difficult to attach insulation and the interior finished wall to the concrete. The stud walls of the AWWF allowed us to hide the wiring and plumbing, too.

The AWWF also suited our construction schedule. We were able to start on it as soon as the excavator finished the cellar hole. For a poured foundation, we would have had to hire a contractor and wait for him to work our job into his schedule. We would then have had to wait for the forms to be erected, and the concrete to be poured and to cure. For a concrete-block foundation, we would have had to pour a concrete footing and then begin the time-consuming task of laying the blocks.

The AWWF facilitated the installation of windows and doors. We simply nailed and screwed them in, as in ordinary frame construction. An error in pouring a concrete foundation can be disastrous when it comes to installing doors and windows; just a little reframing corrects an error with the AWWF. We decided to relocate a door slightly—it took us only two hours.

We wanted to use spruce clapboards for our exterior siding, and we would be able to nail the clapboards directly to the exposed portion of the wood foundation. We're not sure what we would have done with the above-grade portions of a concrete foundation.

The clincher for us was the money we saved. We were the labor force for everything but the excavation, standing-seam steel roof and drywall. Like most people who build their own homes, we did not include our labor as part of the cost. While we didn't do a detailed analysis for all the options, we estimated a savings of more than $1,000 over the cost of poured concrete. Materials for our 24-ft. by 32-ft. foundation cost us about $1,850 in August 1980.

Many people ask us if we're worried about the foundation rotting out. We aren't. The required preservative salt retention in the pressure-treated wood is 0.60 pounds per cubic foot, which is 50% higher than building codes require for general ground-contact applications. The USDA Forest Service's *Wood Handbook: Wood as an Engineering Material* indicates that test stakes in Mississippi have lasted more than 20 years at lower preservative salt retentions.

The AWWF is a stud wall sheathed with plywood on the outside. The walls stand on footing plates, which lie on a pad of gravel. A concrete slab poured inside the walls prevents the backfill from pushing in the bottoms of the walls. The tops of the foundation walls are securely fastened to the first-floor structure before the backfilling begins.

The size of the footings and studs and the thickness of the plywood depend on the size of the building, the grade of lumber, the depth of the backfill and the spacing of the studs. An industry booklet, "The All-Weather Wood Foundation: Why, What and How" ($1 from the American Plywood Association, Box 11700, Tacoma, Wash. 98411), supplied enough information for us to design our AWWF with confidence. We used 2x10 footing plates, except at the back wall, where we used 2x8s because of the smaller load on the back wall. Our design called for ½-in. plywood and 2x8 bottom plates, studs and top plates. The top plates and the plywood that was more than 1 ft. above grade were not preservative treated. We choose to use 2x8 studs on 16-in. centers so we could fit two layers of R-13 insulation inside the exterior walls. Our house design required very low-grade studs (F_b ot 975 psi minimum) for adequate strength with our depth of backfill (up to 5 ft.). The fasteners were 10d stainless steel nails to connect the bottom plates to the studs and footing plates, 8d stainless steel nails to connect the treated plywood sheathing to the studs below grade, 8d hot-dipped galvanized nails to connect the untreated plywood to the studs above grade, and

Once the gravel pad is compacted and carefully leveled, left, the footing plates are set in place around its perimeter. The layout of the excavation and the drainage pipes is shown in the drawing on the facing page. Right, pressure-treated foundation framing is built 8 ft. at a time and tilted up. The interior bearing wall is framed with less costly untreated lumber because it won't be in contact with water or wet earth. Photos: Irwin and Diane Post.

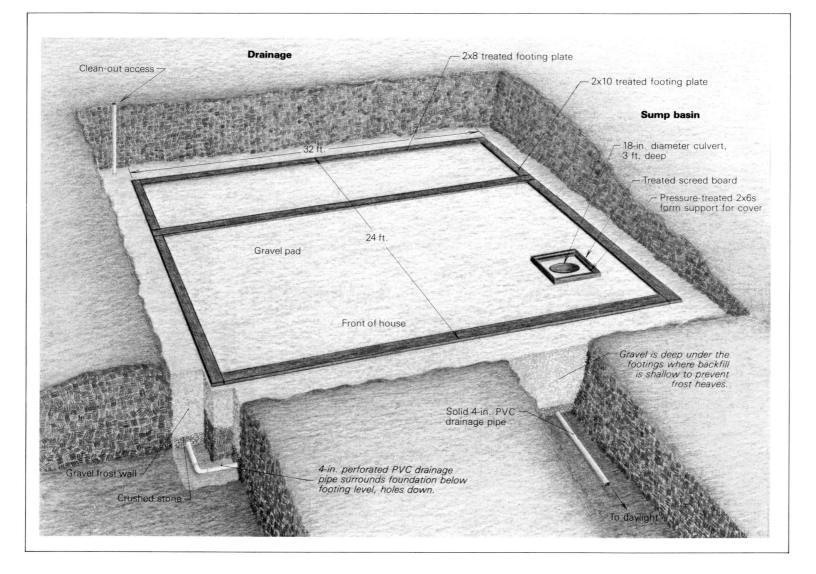

Drainage

Clean-out access

2x8 treated footing plate

2x10 treated footing plate

Sump basin

18-in. diameter culvert, 3 ft. deep

Treated screed board

Pressure-treated 2x6s form support for cover

32 ft.

24 ft.

Gravel pad

Front of house

Gravel is deep under the footings where backfill is shallow to prevent frost heaves.

Solid 4-in. PVC drainage pipe

Gravel frost wall

4-in. perforated PVC drainage pipe surrounds foundation below footing level, holes down.

Crushed stone

To daylight

16d hot-dipped galvanized nails to connect the top plate to the studs.

We found two lumberyards that were willing to bid on our AWWF materials. The better bid was for far higher-quality studs than we needed and for the 5⁄8-in. plywood the supplier had in stock rather than the 1⁄2-in. plywood we'd specified. We could have saved if we had been willing to wait for material closer to our specifications. The price for the stainless steel nails seemed unbelievable at $6 per pound. This works out to about 10½ᶜ for each 10d nail!

In ordering our AWWF materials, we specified that the wood had to be stamped with the American Wood Preservers Bureau (AWPB) foundation grademark, which ensures that the wood is properly treated for use in foundations. We were pleased to find many plugged holes in our material where samples of wood had been removed after treatment to check for retention of the preservative.

Excavation and drainage—We had two excavators bid on our job. Neither one had ever worked on a cellar hole for an AWWF. After reviewing our engineering drawings and instructions, one excavator seemed reluctant. The other showed interest, so we chose him.

The main objectives of the excavation were to lay the gravel pad on which the footings sit and

to provide good drainage around the foundation. Good drainage is an absolute necessity for any foundation. After having suffered with wet basements, we were not about to take any shortcuts with our new house. The design we settled on is diagrammed above.

In the front and sides of the cellar hole, where the footings were going to be less than 4 ft. below finished grade, we wanted a frost wall built to prevent frost heaving under the footings. We had a ditch 2 ft. wide dug to about 5 ft. below finished grade, and set drainage pipes surrounded by crushed stone in the bottom. The ditch was then backfilled with gravel.

Along with the frost wall, we had a sump basin excavated inside the foundation walls. This basin is simply a hole in the basement floor connected by pipe to the drains around the house. Groundwater normally flows to the drain-pipe outlet. If the outlet becomes plugged, the water backs up into the sump basin so we can pump it outside, keeping our basement dry. We used a 3-ft. section of corrugated aluminum culvert 1½ ft. in diameter to form the sump basin. A treated wood cover over its top is flush with the concrete slab.

Next, we worked on the gravel pad. It varied in depth from 6 in. (the required minimum) to more than a foot. We used crusher-run gravel, which does not contain large cobbles. Pea-stone,

coarse sand or crushed rock could have been used instead. The excavator drove his bulldozer back and forth over the gravel to compact it. We made sure the pad was large enough to lay out our footings at 24 ft. by 32 ft., and we marked the locations for the footing plates.

After the gravel was compacted, we leveled the pad by driving 2x2 stakes of treated wood into the gravel at 6-ft. intervals along the critical lines where the footing plates were to lie. Using a surveyor's level for accuracy, we drove the top of each stake to exactly the same elevation, which was close to the average elevation of the gravel. Then, using shovels and garden rakes, we leveled the entire pad to match the tops of the stakes. To level the footing lanes exactly, we scraped a straight 2x4 stud over the gravel between the stakes. Leveling took half a day and some patience and care.

Next we installed the ABS wastewater pipes that were to run under the basement floor. We laid out their positions precisely on the gravel surface, then we dug the ditches, put the pipes in place (gluing the joints well), covered them, and releveled the disturbed areas. We were then ready to set the footings.

Framing—The footing plates were simply laid flat on the ground (photo, previous page, left). We made careful diagonal measurements with a

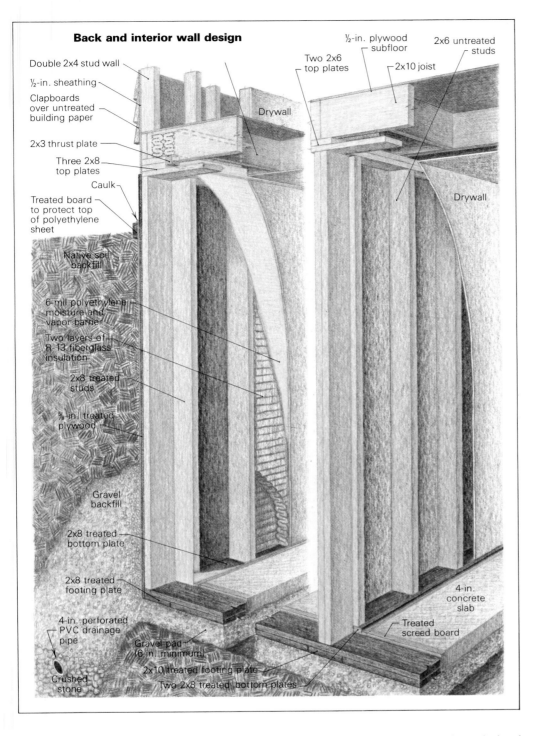

Back and interior wall design

Double 2x4 stud wall

½-in. sheathing

Clapboards over untreated building paper

2x3 thrust plate

Three 2x8 top plates

Caulk

Treated board to protect top of polyethylene sheet

Native soil backfill

6-mil polyethylene moisture and vapor barrier

Two layers of R-13 fiberglass insulation

2x8 treated studs

⅝-in. treated plywood

Gravel backfill

2x8 treated bottom plate

2x8 treated footing plate

4-in. perforated PVC drainage pipe

Crushed stone

Gravel pad (6-in. minimum)

2x10 treated footing plate

Two 2x8 treated bottom plates

½-in. plywood subfloor

Two 2x6 top plates

2x10 joist

2x6 untreated studs

Drywall

Drywall

4-in. concrete slab

Treated screed board

fiberglass tape to ensure that the footings were positioned squarely. Sections were cut out of the plates to accommodate the wastewater pipes located in the bearing walls.

There were just the two of us, so we were not able to handle long, heavy wall sections. We framed one 8-ft. section at a time, stood it in place, nailed it to the footing plates, and nailed on the top plates so as to connect adjacent wall sections. As the wall took shape, we frequently checked for plumb with a 4-ft. carpenter's level and a plumb bob.

The drawing above contrasts the design differences between the back wall and the interior bearing wall in our foundation. We did not use preservative-treated wood for the interior foundation bearing wall of 2x6 studs. By adding an extra bottom plate of treated wood and trimming off the appropriate length from each stud,

we raised the bottom of the studs above the level of the concrete slab. Using untreated studs in this wall saved us a lot of money.

When the foundation walls were all in position, we nailed sheathing onto their lower halves and fully sheathed some of the corners to stiffen the structure. Then we brushed the cut ends of the foundation wood with a generous coat of preservative, and applied a bead of silicone caulking between every sheet of plywood on the foundation.

We attached the first floor joists to the top of the foundation so the floor structure would resist the force of the backfill against the walls of the foundation. To make an especially strong connection at the back of our house, where the fill is deepest, we nailed an extra top plate and a 2x3 thrust plate onto the back foundation wall, and notched the joists to fit. To stiffen the end walls,

where the fill is deep, we added blocks between the two outer joists at 4-ft. intervals. For additional strength we glued (as well as nailed) the floor decks onto the joists.

It took us six days to erect the foundation walls, sheath their lower halves, attach the first floor joists and deck the first floor. We have read that experienced crews working with a small crane and prefabricated wall sections can completely erect AWWF walls in a few hours.

Finishing up—The concrete slab was poured after the first floor deck was completed. We prepared the floor by re-leveling the gravel in the foundation. This did not require as much accuracy as leveling for the footings—half an inch tolerance was acceptable. We shoveled excess gravel outside and laid a 6-mil polyethylene sheet on the gravel and a few inches up the walls as a moisture barrier. We also nailed screed boards of 1x3 treated wood around the sump basin and along the long sides of the two floor sections (one covering the front two-thirds of the foundation, and the other over the back third). Besides providing guidance in spreading the wet concrete, the screed boards helped hold the plastic in position.

The concrete truck pulled up to the front of our house, and the chute was put through the large window openings. We used a homemade chute extension, built from plywood and 2x10s, to reach the back third of the house. Aside from the person who delivered the concrete, we had one other to help us with the pour. The resulting slab was roughly 4 in. thick, with its surface about 1 in. above the bottom of the exterior studs. The openings between the studs provided plenty of air circulation so we used a gasoline-powered trowel for surface finishing.

After the slab was in, we finished the foundation sheathing. The excavator completed the drainage-pipe loop around the back of the house. The pipe was set lower than the footing plates all the way around. As in the bottom of the frost wall, it was surrounded with crushed stone. We had a cleanout installed in the highest section of the loop in case we ever need to flush the drainage system.

Next, we draped 6-mil polyethylene sheeting around the foundation from finished grade to just below the bottom of the footing plate, and protected its top with a 1x4 strip of treated wood caulked along its top edge. Gravel was used as backfill close to the foundation, and the finished grading included sloping all the surfaces away from the house to direct surface runoff away from the foundation.

The basement in the house we built has turned out to be very warm and dry—a very comfortable living space. The ease and speed of building the AWWF was outstanding, and the cost savings over the other types of foundations was significant. Our experience makes us wonder how long it will be until All Weather Wood Foundations displace concrete foundations, just as concrete foundations have displaced those of fieldstone.   ☐

Irwin and Diane Post are forest engineers. They live in Barnard, Vt.

Capping a Foundation

One man's method for raising wood sills
built too close to the ground

by Roger Allen

Many older homes were constructed with wooden sill plates too close to the damp ground. The primary reason for capping a foundation is to correct this situation by raising the sill. While this isn't the sort of task most homeowners would consider doing themselves, it is not as difficult as one might think. If the ground around the house can't be lowered, then with proper planning and careful workmanship, it's usually possible to raise the foundation.

In capping, siding is removed to expose the studs, and while the house is supported with temporary beams or house jacks, the studs are shortened to accommodate the cap. Then a new foundation is poured on top of (and in some cases, around) an existing foundation (drawing facing page, top.) Minimum clearances between sill and grade allowed by building codes vary, but 6 in. to 8 in. is the general rule. Reinforcing steel (rebar) and anchor bolts are included as in any other foundation. The new anchor bolts are an additional benefit of capping since many older homes were constructed without them.

The type of reinforcement used depends on whether the existing foundation is brick or concrete. In capping a concrete foundation, horizontal rebar and vertical dowels are added into the existing foundation, as shown on the facing page, top right. The size, amount and placement of this steel should be determined by an engineer. To cap a brick foundation, a saddle cap is constructed to strengthen the brick. This method requires the cap to encompass the brick foundation with a minimum of 3 in. of concrete on the sides. The height is determined by the necessary clearance above grade. Horizontal rebar is added to each side and to the top. In high caps a second horizontal piece may be required on top. A saddle tie, which is a piece of bent reinforcing steel, joins the horizontal pieces. The saddle tie is placed at approximately the same distance required of anchor bolts (generally every 4 ft.).

Capping a brick foundation has additional benefits. Old bricks and mortar tend to be more absorbent than new concrete, and capping prevents the bricks from acting as a wick, absorbing moisture from the ground and transferring it to the framing. It also adds reinforcing steel where there was none, and anchors the house to a solid wall of concrete.

Getting the house off its foundation—Moving the structure poses obvious problems. How do you pour concrete on top of a concrete wall that supports the house? How do you support the house when necessary framing is removed? And how do you pour concrete into that dark, underground world of spiders and snails, where a human can barely crawl, let alone work?

If the challenge of working under such conditions does not entice you, think of the money you will be saving. To cap 120 ft. of a foundation with an 8-in. wide and 12-in. high cap, you'll need about 3 cu. yd. of concrete. At a cost of $55 per cu. yd., that means $165 for concrete. You may need $100 more for a concrete pump, and another $100 for reinforcing steel, anchor bolts and miscellaneous hardware. Forming lumber could run yet another $100, and one or two helpers on the day of the pour perhaps $100 more. If you consult an engineer, it could cost you another $100. All this adds up to $665. I have seen estimates as high as $8,000 for the same job. Perhaps spiders aren't so scary.

Because the existing sill must be removed where the foundation is to be capped, temporary supports must be constructed. It is wise to cap a foundation in several sections and avoid supporting the entire house at once. I generally study how the house is normally supported and duplicate this support as closely as possible. This is one case when you should always overbuild your bracing. If you have any doubt as to how strong the temporary supports should be, consult an engineer or an experienced builder.

Support procedures—Where the floor joists run perpendicular to the foundation, a beam (usually a 4x8) can be run underneath them a few feet in from the foundation (figure 1A). The posts (one post every 8 ft. is usually enough) should be set on a solid pad in undisturbed soil. The pads must be large enough to distribute the load over the ground without sinking. On very solid earth, precast concrete piers make excellent pads. Heavy blocks of wood or timbers will often suffice, but in some cases I have had to dig a hole 4 in. deep by about 18 in. square to pour a concrete pad.

House jacks and shim shingles can simplify the installation of temporary supports. When the old framing is removed, the house will settle a little. To minimize settling, I aim for a snug fit of the temporary support posts. House jacks can be used in place of or in conjunction with the posts to snug the beam up to the floor joists. Shim shingles tapped between the posts and beam also help to ensure a tight fit. Support posts should be plumb and checked periodically for leaning. Be sure that all the posts and beams are secured strongly enough with diagonal bracing to withstand collisions from workers.

When the joists run parallel to the foundation, another method of support is called for. In this case a rim joist will be carrying the load when the studs beneath are removed or shortened. If the rim joist is not doubled, it should be at this time; to prevent tilting, add blocking to the adjacent joist (figure 1B). The distance between the temporary support posts will be determined by the allowable span of the joist. If the joist is doubled and nailed securely, it will increase the allowable span to about 6 ft. in a standard platform-framed, two-story house. When the house is supported this way, repairs must be made in sections between the supports.

If you want to repair the whole wall at one time, place 4x8 beams every 6 ft. perpendicular to the rim joist and support them at either end, as in the drawing on the facing page, bottom right. This technique allows longer spans of new sill to be installed at one time, but you have to remove more exterior siding than may be necessary, and you have to work around twice as much support timber. Nevertheless, it may be the best or the only technique for a particular job. Don't feel limited to only one technique for supporting a house. Many times a combination works well.

Removing existing sills—Once the house is supported, cut the existing studs to the proper height to accommodate the raised foundation. It is often easier to remove the exterior siding and work from the outside, but before choosing this approach, consider that exterior siding left in place makes an excellent concrete form; stucco also works well. On a concrete foundation, if there is adequate work space on the inside, leave the siding on and remove what is necessary after pouring the concrete. On a brick foundation, this consideration is irrelevant because the existing foundation must be encompassed on both sides.

The old studs must be cut off very straight. When the siding is left, a Sawzall is generally the tool to use, but making a straight and square cut with a Sawzall is difficult. Skilsaws cut square but in the cramped quarters you may be working in, Skilsaws are unwieldy, often dangerously so. A good sharp handsaw (and some elbow grease) is often the best choice.

Before cutting the studs, connect them with a well-nailed tie brace to keep the studs from moving once they lose the connection with the old

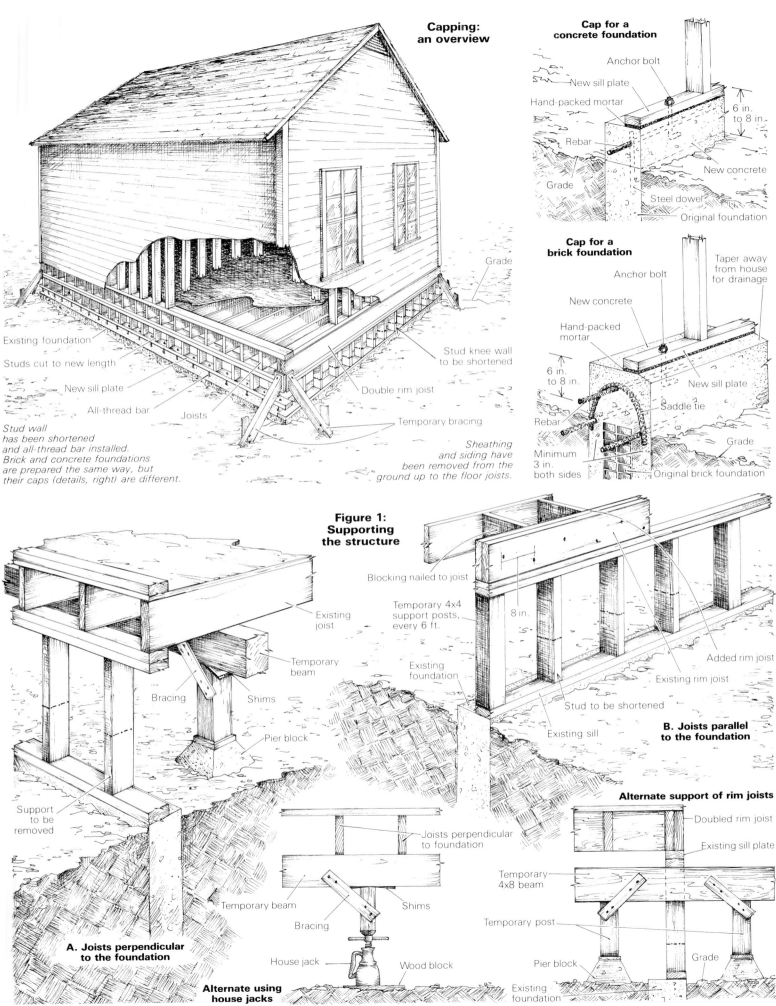

Capping: an overview

Cap for a concrete foundation

Anchor bolt

New sill plate

Hand-packed mortar

6 in. to 8 in.

Rebar

New concrete

Grade

Steel dowel

Original foundation

Existing foundation

Studs cut to new length

New sill plate

All-thread bar

Joists

Stud wall has been shortened and all-thread bar installed. Brick and concrete foundations are prepared the same way, but their caps (details, right) are different.

Grade

Stud knee wall to be shortened

Double rim joist

Temporary bracing

Sheathing and siding have been removed from the ground up to the floor joists.

Cap for a brick foundation

Anchor bolt

New concrete

Hand-packed mortar

6 in. to 8 in.

Rebar

Minimum 3 in. both sides

Taper away from house for drainage

New sill plate

Saddle tie

Grade

Original brick foundation

Figure 1: Supporting the structure

Blocking nailed to joist

Temporary 4x4 support posts, every 6 ft.

8 in.

Added rim joist

Existing foundation

Existing rim joist

Stud to be shortened

Existing sill

B. Joists parallel to the foundation

Existing joist

Temporary beam

Bracing

Shims

Pier block

Support to be removed

Alternate support of rim joists

Doubled rim joist

Existing sill plate

Temporary 4x8 beam

Temporary post

Pier block

Existing foundation

Grade

Joists perpendicular to foundation

Temporary beam

Bracing

House jack

Shims

Wood block

A. Joists perpendicular to the foundation

Alternate using house jacks

Illustrations: Barbara Smolover

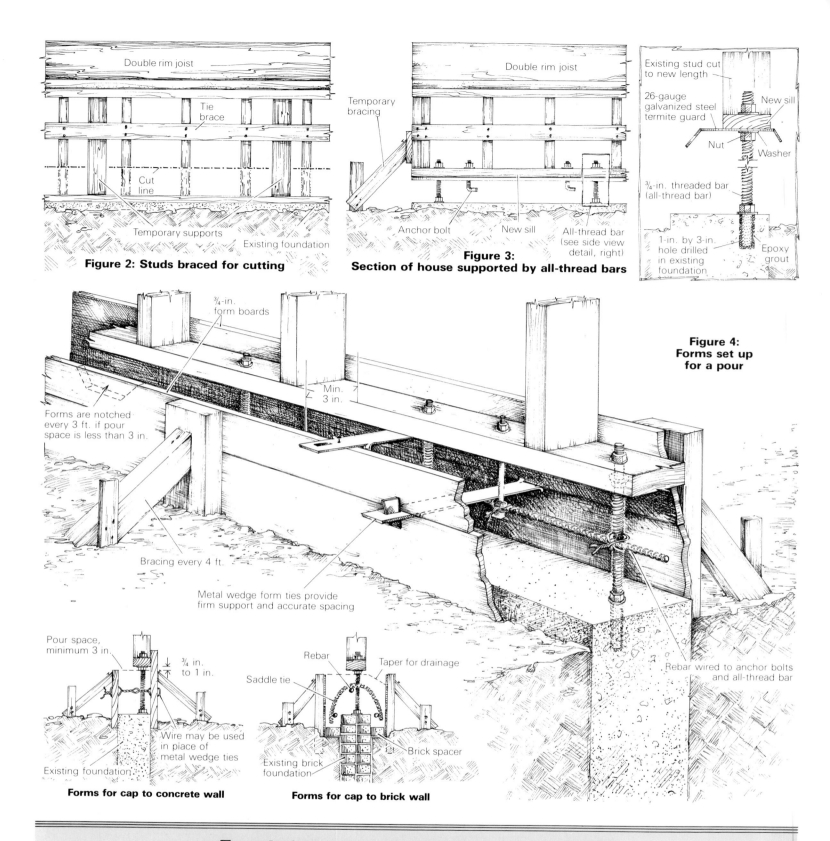

Double rim joist

Tie brace

Cut line

Temporary supports

Existing foundation

Figure 2: Studs braced for cutting

Double rim joist

Temporary bracing

Anchor bolt

New sill

All-thread bar (see side view detail, right)

Figure 3:
Section of house supported by all-thread bars

Existing stud cut to new length

26-gauge galvanized steel termite guard

New sill

Nut

Washer

¾-in. threaded bar (all-thread bar)

1-in. by 3-in. hole drilled in existing foundation

Epoxy grout

Figure 4:
Forms set up for a pour

¾-in. form boards

Min. 3 in.

Forms are notched every 3 ft. if pour space is less than 3 in.

Bracing every 4 ft.

Metal wedge form ties provide firm support and accurate spacing

Rebar wired to anchor bolts and all-thread bar

Pour space, minimum 3 in.

¾ in. to 1 in.

Wire may be used in place of metal wedge ties

Existing foundation

Forms for cap to concrete wall

Rebar

Taper for drainage

Saddle tie

Brick spacer

Existing brick foundation

Forms for cap to brick wall

Foundation capping: a step-by-step summary

1. Note the presence of water lines, sewer lines, meter boxes, wiring, and other obstacles that may affect the placement of temporary and permanent supports.
2. Support the house where capping is to be undertaken.
3. Nail a tie brace to studs above cut line.
4. Remove the old sill so your sawblade won't bind when cutting the old studs.
5. Cut old studs to proper height.

6. Install anchor bolts and all-thread bar to a pressure-treated sill.
7. Nail new sill to bottom of cut studs.
8. Drill holes in old foundation for all-thread and rebar if needed. (This may precede step 7 if workspace is limited.)
9. Grout all-thread into old foundation and tighten the nuts on the all-thread between the old foundation and the new sill. Remove temporary supports.

10. Hang rebar.
11. Build forms.
12. Pour concrete.
13. Vibrate.
14. Allow concrete to cure.
15. Hand pack mortar under new sill.
16. Remove forms.
17. Tighten anchor bolts.
18. Remove temporary bracing.
19. Replace siding if it has been removed.

sill. This step, shown in figure 2 on the facing page, is especially important where the siding has been removed.

Installing a new sill—Once the studs have been cut and the lower parts removed, nail the new sill to their bottoms. Select a good straight pressure-treated sill. When the sill is cut to expose a cross section, the lumber should be green all the way through. If you see blond lumber, the pressure treatment was not thorough. Use only well-treated sills—you won't want to do this job a second time.

It may be easier to drill the holes for the anchor bolts before the new sill is nailed to the bottom of the studs. Take care not to place a bolt where a stud will fall. The size and spacing of anchor bolts depend on local requirements. In areas prone to earthquakes, anchor bolts are commonly placed 4 ft. apart, and no more than 1 ft. from corners, ends, or unions in the sill. Elsewhere, 6-ft. spacing is the norm. Common anchor bolts range in size from $\frac{1}{2}$ in. to $\frac{5}{8}$ in. by 8 in. to 10 in.; $\frac{1}{2}$ in. by 10 in. is typical.

If termite-proofing sheet metal is to be added, fasten it tightly to the bottom of the sill before drilling and nailing. Drill through the sheet metal first using a hole saw attachment. Yes, this does wear the sawblades out fast.

If large sections of foundation are to be capped in one pour, all-thread bar should be installed. All-thread bar, or threaded steel rod, is metal stock threaded its entire length. It is available in various sizes from hardware stores or concrete accessory companies. At the same intervals that one would place a support post, the all-thread bar should be run from a 1-in. diameter hole drilled 3 in. deep into the old foundation through the top of the new sill, as shown in figure 3. Pack grout or an epoxy equivalent into the gap between the $\frac{3}{4}$-in. rod and the larger hole to ensure a firm connection with the old concrete. Place nuts and heavy washers at the bottom of the bar on top of the old foundation and directly under the new sill. When the nuts are tightened, the all-thread will secure the nailed sill firmly to the framing above and prevent it from sagging. When they are properly installed, the all-thread bars support the house and the temporary posts can be removed.

All-thread bar must be installed plumb or it will not carry the house properly. Although the all-thread bar is adequate to support the house in short sections, it does not supply any shear strength; it only carries a load from directly above. It is also necessary to attach temporary diagonal bracing to the corners of the house, as shown in figure 3.

If there is enough room above the new sill, the all-thread can be added after the sill is nailed in place. Otherwise insert it into the sill before installation. Holes in the softer brick foundations may be drilled with masonry bits and a normal electric drill. For a concrete foundation, considering that many holes have to be drilled, you can rent a Roto-hammer—a combination jackhammer and drill—from most equipment yards. Remember to drill the holes in the sill and foundation $\frac{1}{4}$ in. larger than the bolts to allow some leeway for installation.

Forming the cap—When the new sill is nailed to the bottom of the shortened studs and the all-thread is in place and tightened, it is time to install the rebar. The all-thread is a handy place from which to hang it. If the wall is to be poured in sections, the steel should extend 24 in. beyond to provide an effective tie to the next section when it is poured. If vertical steel is used, it should be inserted into holes at least 3 in. deep in the old foundation and grouted.

Because not much concrete is usually involved in pouring a cap, forming is relatively simple. Often all you have to do is to brace boards that are nailed in place to wooden or metal stakes. If the exterior siding is left in place as a form, make sure that weak points, such as the union of two pieces of siding or a crack in the stucco, are strong enough. If not, beef them up.

When the cap is to lie only on top of the old foundation and not encompass it, the outside form will rest tightly against the old foundation's side. If the siding has been removed, these form boards can be nailed with duplex nails to the framing above. Duplex nails should be used on all forms to allow for easy removal. The inside forms will also rest against the old foundation and require wood or metal stakes and form ties or wire for proper support (figure 4).

When the cap rests only on top of the foundation, leave space at the top of the inside form for pouring the concrete. Usually the size of the sill in relation to the size of the wall will allow for this space. If more space is required (3 in. is the minimum) notch the form at 3-ft. intervals at the top and build a funnel for the concrete to fall into the notch. If a termite guard has been added, you can temporarily bend the edge upward and out of the way before pouring the concrete.

A cap for a brick foundation requires a minimum 3-in. spacer every 4 ft. at the bottom of the form. Bricks work well for this purpose. After the concrete is in place and has been vibrated, use a small trowel to taper the exterior top away from the house for drainage.

Many people assume that because a cap will hardly be seen there is little reason to make it look good from within. Consequently, there is a tendency to build sloppy-looking forms on the interior side. But let me caution you: Sloppy-looking forms are often weak. Straight, neat forms are much easier to brace properly.

Pouring concrete—Most of the hard work is now over and the most exciting part is about to begin: the pour itself.

For small sections of foundation, sacks of concrete can be mixed in a wheelbarrow and distributed via shovel or bucket. For larger amounts, a ready-mix truck should deliver your calculated amount. Always add at least 10% to the estimated load to allow for the inevitable tipped wheelbarrow. If the volume of concrete is very large or the workspace too awkward to drag buckets of concrete, you'll have to hire a concrete pump. Hire a grout pump and use concrete made with pea gravel, a concrete mix with aggregate no larger than $\frac{3}{8}$ in. The mix costs about $10 more per yd., but it allows you to use a smaller and less expensive concrete pump—the grout pump. A grout pump has the added advantage of smaller hoses, which means less weight to drag around under the house.

Concrete pumps and concrete are usually ordered from separate companies—plan ahead to synchronize the two. Even so, on the day of the pour, one is likely to be late. Don't panic, it always happens. It is far better, however, to have the pump arrive early so it can be set up before the concrete arrives.

Before you begin to pour, be sure that the operator of the pump understands your instructions as to starting and stopping the pump. Learn early what the delay between your instruction and his response will be. If you are far under the house, another person can relay your commands. I can recall several times I was almost buried in overflowing concrete because a pump operator was talking to a neighbor or eating his lunch. If you have calculated the volume of concrete as closely as you should have, you don't want it wasted.

A critical aspect of pouring is to make sure that no spaces are left unfilled in the forms. The concrete must be vibrated so it settles everywhere. Vibrating guards against the weakening effect of a honeycombed wall. There are several methods of vibrating. An electrical vibrator, designed specifically to be placed into the concrete on large pours, is rarely necessary for capping. One good technique for capping is to tap the sides of the forms with a hammer, as well as packing the concrete from the top with a stick. Another way is to remove the blade from a Sawzall and place the shoe of the Sawzall against the outside of the form; when running, the Sawzall acts as an excellent vibrator.

Vibrating increases the stress that the poured concrete puts on the forms, so the forms should be watched as carefully when vibrating as they were during the pour. If the bracing seems inadequate, stop pouring or vibrating and shore up the weak sections immediately.

Because concrete shrinks as it dries, a small gap would be left under the new sill if the concrete were poured all the way to the bottom of the sill and then allowed to dry. To solve this problem, leave a $\frac{3}{4}$-in. to 1-in. gap between the newly poured concrete and the bottom of the new sill. After the concrete has cured, you can hand-pack mortar into this space. If the gap were smaller, filling would be more difficult.

All concrete spilled onto the forms and stakes should be cleaned off before it hardens, making it easier to strip the forms. Many a worker has cursed his lack of foresight while struggling to pull stakes or to pry boards that are embedded in hardened concrete. You can pull off the forms the day after the pour, but I usually leave the temporary supports in place for a couple of days to allow the concrete to cure and attain its full strength before accepting the weight of the house. After the concrete has cured and the hand-packed mortar has hardened, the anchor bolts should be tightened. If siding was left on the outside as a form, it should now be raised to the proper level above grade. Your house now sits high and dry. □

Roger Allen is a general contractor in the San Francisco area.

Cracked Foundations
Prevention is easier than repair

by Kip Park

Throughout much of North America, home-owners face a similar problem—the cracking concrete foundations supporting their houses. The results are rarely life-threatening, but can be plenty uncomfortable. Water seeps into basements, plaster in the house cracks and doors jam.

There's no doubt that the problem is widespread. A 1981 survey by Owens-Corning Fiberglas Corporation of over 31,000 families throughout the U. S. found 59% of the households reporting leaky basements. A similar survey here in Canada in 1983 (by the University of Manitoba civil engineering department) found that over 70% of the houses surveyed had cracks in their basement floors, interior walls and ceilings, with major cracks in 9% of the houses (see photo below). Cracks in foun-

dations are a particular problem in areas where there are expansive soils. Poor construction practices can also cause major problems with cracking.

Unruly substrates—Expansive soils are found throughout North America, but are particularly common here in Winnipeg, which is located on a former glacial lake bottom. Lake Agassiz was formed during the last Ice Age and covered about 180,000 sq. mi. of what is now Manitoba, Saskatchewan, North Dakota and Minnesota. When the lake vanished about 7,000 years ago, a layer of silt and clay which varies in thickness between 15 and 40 ft. was left behind.

Lake Agassiz clay, like most clays, changes in volume as soil moisture levels change. Ac-

cording to Prof. Len Domaschuk of the University of Manitoba civil engineering department (who conducted the Winnipeg foundation survey), a layer of clay one foot thick will change about an inch in thickness as it alternates between dehydration and saturation. As it swells, clay can exert over 10,000 lb. of pressure per sq. ft. against a foundation.

A National Research Council of Canada study in Winnipeg between 1962 and 1966 found that the elevation of the ground surface fluctuated by 5 in. during that period, with vertical movements of 3 in. at 5 ft. below grade. A house built on this unruly substrate with its foundation footings 5 ft. below grade will therefore heave or settle up to 3 in. as moisture levels change. The house will sit still only if the moisture level is stabilized.

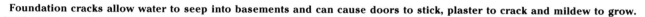

Foundation cracks allow water to seep into basements and can cause doors to stick, plaster to crack and mildew to grow.

Controlling soil moisture—There are some basic and inexpensive measures that can be taken to minimize changes in the moisture content of the soil.

Lots shouldn't be completely cleared of shrubs and trees, as they often are in subdivisions. Roots take moisture from the soil, and when they're removed, soil moisture levels increase and the soil swells. A new concrete foundation placed in this swelling soil will almost always crack, especially if the concrete hasn't reached its full design strength—and often it hasn't.

Conversely, trees shouldn't be planted too near a foundation. They should be kept at least 15 ft. away, otherwise they'll cause the clays against the house to shrink excessively and cause the house to settle. If there's a tree right next to the foundation, remove it.

In Winnipeg, new houses are now required to be "perched" by building up the soil around the house so surface water drains away from the foundation. The house must be perched a minimum of 3 in. above grade, with slopes extending from the house a minimum of 7 ft. at the front and rear and a minimum of 3 ft. in side yards. Also, the entire lot must be sloped—typically about 6 in. from front to rear on a 100-ft. deep lot—to direct surface water off the lot. Skew the house at an angle to the slope so it won't dam up the runoff, or slope the grade from the center out to the corners of the house to keep water from puddling against the foundation. Around existing homes, the soil will settle over time, creating a natural ditch which directs runoff water toward the foundation. Inspect the level of the soil regularly and add fill as needed to maintain a slope away from the house.

Downspouts direct rainwater away from the building. Splashpads are a must. The most durable and maintenance-free of these are made of precast concrete. They should be at least 3 ft. in length to prevent rainwater from seeping down the side of the foundation and into the expansive soil. Flexible plastic tubes attached to downspout ends help to carry rainwater away from the house.

Sidewalks, where they abut a foundation, should be sloped to direct water away from the house. When replacing a sidewalk, remove the deteriorated concrete. Don't place the new sidewalk over the old one, because the additional weight puts more pressure on the foundation and can produce horizontal cracks in the foundation wall. Concrete driveways should be at least 5 ft. away from foundation walls for the same reason.

Design and placement of concrete—It's not just expansive soil that causes foundation cracking. Concrete that is poorly designed or improperly placed or cured will add to the problem. In Manitoba, poured concrete, ideally reinforced with rebar, is nearly always used for residential foundations. It provides a stronger barrier than concrete block against expansive soil and is the cheaper alternative in our area. Concrete is subject to a lot of abuse

when it's being placed—and note the distinction between *placing* concrete and *pouring* concrete. Pouring concrete suggests that it has the consistency of soup. This occurs when water is added to the mix on site, as workers often do in order to make the concrete easier to move around in the forms. But adding even one gallon of water to a cubic yard of concrete lowers its compressive strength by more than 150 lb. per sq. in. (psi). The watered concrete is prone to greater shrinkage and cracking during setting, and its durability and watertightness is reduced.

Canada's National Building Code allows a maximum slump of 4 in. in concrete used for house foundations. That's pretty stiff, and the concrete isn't easy to move around. To make the job easier, place the concrete in several different locations around the perimeter of the form. Or use chutes, buckets, or a pump truck to get the concrete as close as possible to its final destination. You might want to ask the concrete plant to add a superplasticizer, which helps prevent segregation of the concrete while improving its workability. Don't drop the concrete more than about 6 ft., as this will cause the segregation of aggregates, fines and cement, which decreases the strength and watertightness of the wall. Don't vibrate the concrete more than necessary, and don't move it with a rake either, as both techniques will cause segregation.

Mixed with the concrete at the batch plant, additives called air-entraining agents create a system of tiny bubbles throughout the mix. In addition to improving the workability of the concrete, the bubbles create a void into which moisture within the concrete can expand during freezing, thus preventing damage to the concrete itself. The Canadian Standards Association (CSA) requires 3% to 6% air entrainment for footings and 5% to 8% for walls, garage slabs and other exterior concrete. Ready-mix plants in Canada sometimes include an air-entrainment admixture at no extra charge.

There's another additive that has only recently hit the market. Fibermesh (Fibermesh, Inc., 4019 Industry Dr., Chattanooga, Tenn. 37416) is claimed to help prevent shrinkage cracks in the concrete, which typically develop within 40 days of the concrete placement.

Fibermesh consists of millions of individual ¾-in. to 2 in.-long interconnected polypropylene fibers, which are uniformly distributed throughout the concrete. They are dumped in the mixing drum on site at the rate of 1 to 1½ lb. per cu. yd. of concrete.

As concrete hardens, microscopic cracks develop as the result of plastic and drying shrinkage. When these cracks intersect the nearest Fibermesh strand, they are blocked and prevented from developing into larger cracks. Fibermesh retards concrete setting by about half an hour because the fibers tend to hold moisture in the mix, but this can be an advantage. In the Winnipeg area, it costs $20 per cu. m. ($13 per cu. yd. U. S.) to add Fibermesh to concrete.

Curing—Concrete will reach its full design strength only if it is allowed to cure in conditions that prevent the rapid loss of moisture. Concrete forms should be moistened with water or oiled, and the subgrade should be dampened to prevent water from being sucked out of the concrete. If you oil the forms, apply the oil before assembly so it doesn't end up on the subgrade, footings or reinforcing steel. The CSA requires forms to be left on the concrete for a minimum of 24 hrs. after placement, but recommends leaving the forms for the walls in place for at least two days to prevent the concrete from drying out too fast.

Erecting a concrete foundation in extreme temperatures can also affect its durability. When temperatures are above 85° F, concrete should be placed in the early morning or in the evening to reduce the loss of water through evaporation, which reduces the strength of the concrete. When ordering the concrete, ask the dispatcher to reduce the temperature of the concrete, or to add a set-retarder to the mix. This will buy some time to get the concrete in place before it loses too much moisture. Protect fresh concrete from rapid drying and direct sun by covering it with clear polyethylene sheeting between finishing operations. Keep the concrete covered after placement for at least as long as its surface is damp.

While hot weather calls for slowing down the curing of concrete, cold weather calls for the reverse. If concrete freezes before it cures, it suffers a permanent loss of strength and watertightness. Concrete temperatures must be maintained above 50° F during its placement and for three days afterwards. Calcium chloride is sometimes added to accelerate the curing, but no more than 2% of the weight of the cement should be added. It's available in pellet, flake and granular forms. Keep in mind that too much calcium chloride may increase shrinkage cracks and scaling, corrode reinforcing steel and darken the concrete. For walls, leave the forms in place for at least three days to take advantage of the heat of hydration. For flatwork, cover the slab with insulation blankets. In Winnipeg, propane heaters are usually used in conjunction with the blankets during the cold season.

Control joints—Some jobs are likely to crack despite careful placing and curing, and shrinkage is usually the culprit. Some builders use control joints to predetermine the location of shrinkage cracks. The idea is that if you can't beat the crack, you might as well plan for it. The Canadian Home Builders Association says that this could reduce foundation callbacks by 90%. In walls, the CSA recommends vertical control joints every 17 ft., starting 10 ft. from the corners. They also suggest control joints at window openings.

To be effective, the depth of the control joints must be one-quarter of the wall thickness. In either case, the joints must be filled with a bond breaker, which will stretch when the concrete cracks, then sealed with a gun-grade, oil-based caulking. A polysulphide caulking such as Ake Vulkem 116 (Mameco International, 4475 East 175th St., Cleveland, Oh. 44128-3599) can also be used. The caulking must be protected from backfill by asphalt paper or exterior insulation.

Cracks in slabs can also be controlled by using control joints, sawcut 6 to 16 hours after placement or created with boards that remain in the slab after the pour. Joints in slabs can also be created with T-shaped plastic crack initiators (J. A. Crawford Company, P. O. Box 1473, Whittier, Calif. 90609). The top of the T is removed after the slab is troweled. Control joints should be spaced 15 ft. to 20 ft. apart in a rectangular grid in basement slabs. Joint depth should be one-quarter of the slab's thickness.

Drainage and backfill—Exterior foundation insulation is a very effective means of reducing heat loss through the basement. During backfilling, the insulation protects any damp-proofing membrane that is applied to the outside of the foundation. It also creates a resilient spacer between foundation walls and expansive subsoils, thereby reducing pressure on the concrete.

In Canada, Fiberglas Canada Inc. (4100 Yonge St., Willowdale, Ontario M2P 2B6) markets Baseclad as an exterior insulation/drainage product for basements of either new or existing houses. Because Baseclad is made up of fibers oriented vertically, moisture penetrates only the surface layer; moisture droplets are then passed from one fiber to another down its surface to the footings below, keeping moisture away from the foundation wall.

Of course, once the water gets to the footings, it needs someplace to go. Drain tiles collect this water and deliver it via leader lines to a sump pit inside the basement (where it is pumped out to a septic tank or sewer) or to a sewer or drain located away from the house.

Foundations can be weakened or cracked by improper backfilling techniques. One of the major causes of cracks is backfilling against "green" concrete, so it's suggested that backfilling be delayed for at least seven days. Walls should be supported laterally by diagonal bracing from the sill to stakes driven in the ground or by floor joists. Start the backfilling at the corners and work your way toward the middle of each wall.

In new construction, it's a good idea to use crushed limestone or a similar material as a backfill around the foundation instead of the usual excavated soil. The stone promotes good drainage, preventing moisture from collecting around the foundation wall and its footings.   □

Kip Park lives in Winnipeg and writes about housing, construction and energy technology. Photos by the author.

Repairing cracks in concrete

Permanently repairing a cracked concrete foundation requires skill and special materials, but it can be done. The best method is epoxy injection, according to Wally Sokoliuk, a foundation repair expert in Winnipeg, Canada, where extensive pockets of expansive soil make foundation cracks a major problem. It's the only method that seals cracks permanently, other than repairing and sealing the foundation from the exterior, a method which often can't be used.

*Epoxy injection—*Sokoliuk's company, Concrete Restoration Services (77 Paquin Road, Winnipeg, Manitoba, R2J 3V9) has been repairing cracked foundations with epoxy injection for the past 17 years, and reports excellent success. The epoxy he uses is manufactured by Dural International Corp. (95 Brook Avenue, Deer Park, N. Y. 11729) and costs about $100C a gallon ($60 per gal. U. S.). There are three different types, and the use of each is dictated by the width of the crack.

Epoxy isn't the easiest substance to work with, and epoxy injection isn't recommended for the novice. The epoxy has a mere 15 to 20 minute pot life (shorter in warm conditions). Anyone who is unfamiliar with the process will find himself with a pot of expensive hardened epoxy and a crack which has not been completely sealed.

Repairing cracked concrete foundations by epoxy injection is a three-step process, and the work stretches out over several days. First, the crack is prepared by cleaning out the debris (photos below) and drilling 1½-in. deep holes, 8 in. to 9 in. apart, into the crack. Then polyethylene "ports" are inserted into the holes (copper pipe can be substituted). The ports allow the epoxy to penetrate deep into the crack. Next, an epoxy mortar is troweled over the crack. After it has set, another epoxy product such as Duralpatch, is troweled over it to prevent the epoxy from leaking

out later. If water is running out the crack, Sokoliuk uses Duralcrete instead.

The second step is the actual injection of the epoxy. The epoxy is forced into the lowest port with a special injection gun. When the epoxy begins to leak out of the next highest port, the lower port is plugged and the nozzle moved to the flowing port. This process is continued on up the wall until all the ports have been filled.

The final step is to grind off the ports. This is usually done on the third day, when the epoxy has had a chance to reach full-strength (about 4,000 to 9,000 lb. per sq. in., which is stronger than the concrete of most foundations). Sokoliuk says grinding is preferred to using a cold chisel because a chisel might split the epoxy.

Although the process takes three consecutive days, sealing the crack might take only four hours in actual working time. Sokoliuk charges a flat fee of $400C ($350 U. S.) to repair a moderate crack by epoxy injection. On foundations with more extensive cracking, the charge is $40 a lineal ft. ($35 U. S.) He gives a five-year warranty with each injection job. (You can locate a foundation repair expert in your area by contacting the Poured Concrete Wall Contractors Association of America, 825 E. 64th St., Indianapolis, Ind. 46220, or by looking in the Yellow Pages under Foundation Contractors.)

The excavation method—Sometimes foundation cracks can't be repaired by epoxy injection. This is particularly true for houses built before 1940, when concrete quality was usually poor. According to Sokoliuk, there doesn't have to be a crack—the water sometimes just seeps through the wall. Older basements often require sealing from the exterior, a much more labor-intensive, expensive process.

First, Sokoliuk excavates the entire foundation perimeter down to the footing. If necessary, he replaces the drain tiles and covers them with ¾-in. down limestone or a geotextile cloth. Then he trowels a coating of an asphaltic sealant material onto the concrete and embeds a layer of fiberglass material into it. Another coating of the asphaltic sealant follows, with another layer of fiberglass embedded into it, to be covered with a third and final coating of sealant.

A sheet of 6 mil polyethylene sheeting covers the entire foundation. This is protected by an asphalt-impregnated fiberboard (manufactured by C. P. D. Services, 219 Connie Crescent, Unit 14, Concord, Ont. L4K 1L4). Backfilling follows. Ideally, the material used for backfilling should be crushed limestone to provide better drainage around the foundation.

This method of sealing foundations is expensive, running about $100C ($80 U. S.) a lineal foot, so that the average basement will cost between $8,000C and $15,000C ($6,500 to $12,000 U. S.) to seal, says Sokoliuk. The excavation method is limited by existing conditions around the house, such as sidewalks, driveways and shrubs, or the proximity of a neighboring house, which may hamper equipment movement or prevent excavation entirely. —K. P.

Repairing cracks with epoxy injection. One way to repair cracks in basement walls is to fill them with epoxy, a process that takes three days. On the first day, surface debris is removed from the crack with a cold chisel or rotary hammer (photo, facing page). Holes are then drilled in the crack, 1½ in. deep and 8 in. to 9 in. apart, and polyethylene or copper "ports" are inserted into the concrete. Next, a batch of epoxy mortar is troweled over the crack. When it sets, a commercially available epoxy grout is troweled on top (photo, above left). This process prevents epoxy from leaking out in later steps. On the second day, the crack is filled with epoxy. Starting with the lowest port, the epoxy is injected with a special hand pump until it starts flowing out of the next highest port (photos, top right and below right). The lower port is then plugged, and the pump nozzle is moved to a new port. The process is repeated up the wall until the entire crack has been filled. The final step is to grind the ports flush with the surface of the wall. This is done on the third day, after the epoxy has reached full strength.

Digging a Basement the Hard Way

How a builder excavated for a wine cellar under an existing house

by Scott Publicover

I enjoy all phases of building, and as a consequence I've taken on all sorts of construction jobs. I've been a cabinetmaker and a trim carpenter, I've built curtain drains, retaining walls and complicated foundations. This diverse background put me on the short list of contractors called to bid on a wine-cellar project for some folks in Morgan Hill, a town about 30 miles south of San Jose, California. I'd never built a wine cellar, so I agreed to meet the draftsman from Vralsted, Moore, & Associates, Mike Davis, at the job site to learn more about it firsthand.

I got there a bit early and took the opportunity to walk around the three-story house. I was surprised to find another general contractor in the early stages of adding a wing to the place. Huh? Why not have the guy already on site build the wine cellar? The mystery deepened as I looked over the forms for the new foundation. There wasn't a basement in sight.

Meanwhile, Davis arrived and with a wave of his arm directed me toward the front of the house. Standing by the entry, we said our hellos. Then he explained that the owner, Todd Rothbard, had decided that because the house was already incapacitated by the remodeling work for the addition, he might as well add a wine cellar. There was, however, a complication. Rothbard wanted to put the wine cellar under the foyer—the one we were standing next to—the one with three stories of house over it and an 18-in. crawl space under it.

Furthermore, Davis had already interviewed a couple of reputable contractors. Both declined the project, not wanting to take on the assorted liabilities of holding up a house while carting away the very soil that held it up. It sounded like my kind of job. I decided to take it, even though the existing conditions and construction details were still sketchy. I'd simply figure it out as I went, collaborating with the designers and using common sense.

Pier pressure—Rothbard's house was built on a pier-and-grade beam

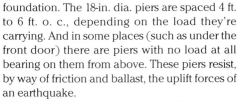

Covered entry. **The author began excavation by erecting a temporary canopy to direct rainwater and pedestrians away from the tractor ramp under the front door.**

Breakthrough. **A Bobcat makes its way under the foundation to carve away the soil for the cellar. The scar on the grade beam below the door shows the former location of a pier.**

foundation. The 18-in. dia. piers are spaced 4 ft. to 6 ft. o. c., depending on the load they're carrying. And in some places (such as under the front door) there are piers with no load at all bearing on them from above. These piers resist, by way of friction and ballast, the uplift forces of an earthquake.

The piers are about 4 ft. deep. They extend through the expansive-clay topsoil to the stable soil below. The grade beams under the foyer cross the crawl space on roughly 15-ft. centers. The wine cellar would fit inside a space defined on three sides by the beams and on the fourth by a line of piers that provided intermediate support to the floor joists.

So just how much foundation can be undermined without risking a cave-in? Like all steel-reinforced foundations, grade beams are just that—beams. And like other kinds of beams, they are capable of spanning gaps without support. I asked engineer George Vralsted how far apart I should space my temporary supports to carry the load of the grade beams. The answer, not surprisingly, was to approximate the spacing of the piers with the spacing of the temporary supports.

Digging under cover—The finished dimensions of the cellar would be 15 ft. square, with an 8-ft. ceiling. Add the depth required for the new foundation and the space needed to work on the block walls, and the excavation came out to 21 ft. square by 12 ft. deep. That added up to over 200 cu. yd. of earth that had to be taken out.

My excavator recommended a model 843, 2½-ton Bobcat for the job. At 5½ ft. wide and 6 ft. high, this little earth mover is compact enough to maneuver in tight spaces. And it has a variety of attachments, such as a clamshell bucket and a backhoe, that make it pretty versatile. But before we could start hauling away the soil, I had to figure out how to get a Bobcat under the house.

It was November, and I didn't want to take any chances on rain

Drawings: Christopher Clapp

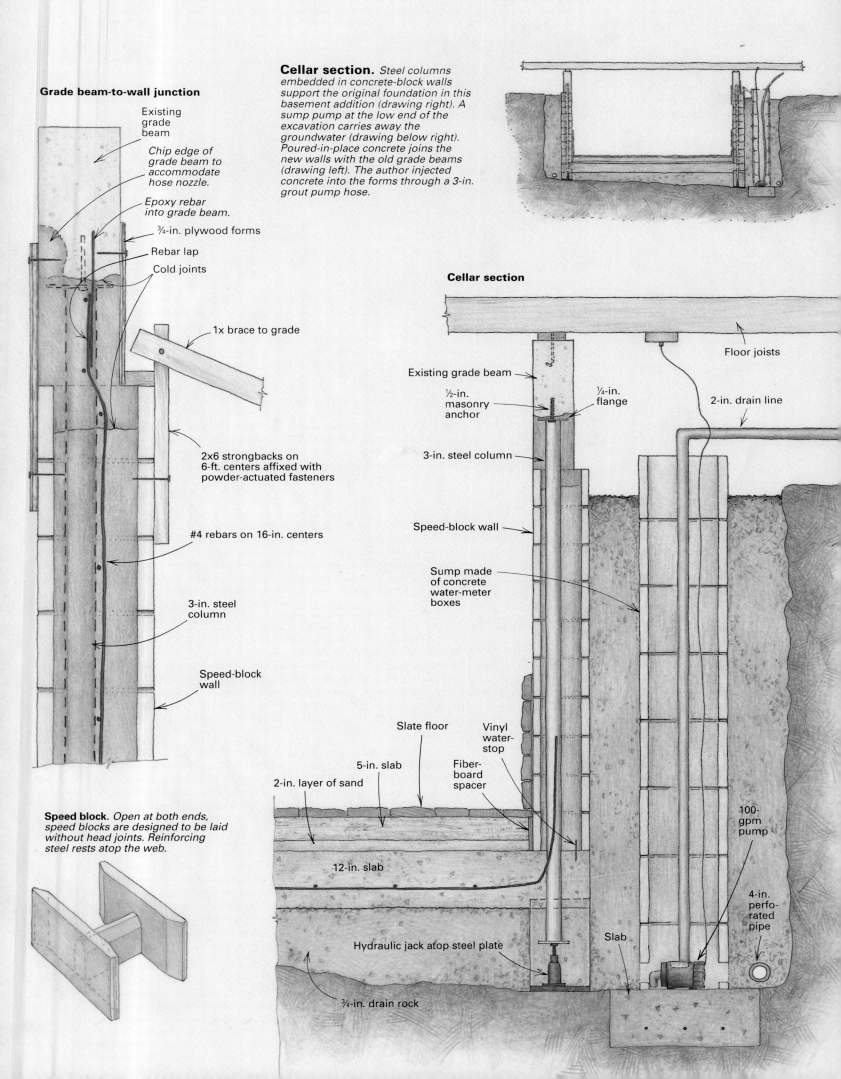

Grade beam-to-wall junction

Existing grade beam

Chip edge of grade beam to accommodate hose nozzle.

Epoxy rebar into grade beam.

¾-in. plywood forms

Rebar lap

Cold joints

1x brace to grade

2x6 strongbacks on 6-ft. centers affixed with powder-actuated fasteners

#4 rebars on 16-in. centers

3-in. steel column

Speed-block wall

Cellar section. *Steel columns embedded in concrete-block walls support the original foundation in this basement addition (drawing right). A sump pump at the low end of the excavation carries away the groundwater (drawing below right). Poured-in-place concrete joins the new walls with the old grade beams (drawing left). The author injected concrete into the forms through a 3-in. grout pump hose.*

Speed block. *Open at both ends, speed blocks are designed to be laid without head joints. Reinforcing steel rests atop the web.*

Cellar section

Existing grade beam

½-in. masonry anchor

¼-in. flange

3-in. steel column

Speed-block wall

Sump made of concrete water-meter boxes

Floor joists

2-in. drain line

Slate floor

5-in. slab

2-in. layer of sand

12-in. slab

Vinyl water-stop

Fiber-board spacer

100-gpm pump

4-in. perforated pipe

Hydraulic jack atop steel plate

Slab

¾-in. drain rock

getting into the excavation. So I started by building a gabled entry off the front door (top photo, p. 68). This temporary shelter would cover a descending ramp to a tunnel under the house. In addition to directing rainfall away from the job, the shelter would act as a barricade to keep people from wandering into the pit. By the time we were directly under the front door, the ramp was 10 ft. deep (bottom photo, p. 96).

Before I built the shelter, my excavator told me the steepest angle that a Bobcat can comfortably climb is about 30°. A ramp with a 10-ft. drop over a 20-ft. run, such as ours, falls within that slope.

Whenever you get a fuel-burning vehicle in an enclosed space, you've got to make air quality a top priority. To make sure the carbon-monoxide levels didn't get dangerous, I rented three squirrel-cage blowers from a rental yard. I put the intakes for two of them outside where they could pick up fresh air, and I placed their outlets in the crawl space. Then I put the third blower in the crawl space where its intake could suck up the heavy carbon-monoxide vapors and convey them outside our tented entry.

Once he had the ramp excavated to the edge of the foundation, my Bobcat man, Von McGee, used the clamshell bucket to scrape away carefully the soil under the grade beam. This would be our method throughout the project for locating the piers. We found one centered under the front door. It was the first to go.

Pier removal—Bobcats have a concrete breaker-bar attachment (typically used for breaking up slabs), and I thought it might be the right tool for busting through the piers. But it delivered such powerful blows that I worried about it cracking the grade beam. Next I tried an electric jackhammer. It weighed an unwieldy 80 lb., and even though I supported it with a sling hung from the joists, it was clumsy to use horizontally. It was also quite slow.

The third tool did the trick. Back at the rental yard I checked out a 40-lb. pneumatic jackhammer and the trailer-drawn compressor required to run it. With a 1-in. chisel bit, this hammer went through an 18-in. pier in about 10 minutes. This is a noisy, dusty job, and even suspended from a sling, the hammer is a handful. I wore a particle mask with a handkerchief over it, goggles, a cap and earplugs. I also placed a blower so that it would direct the dust away from my face.

Severed from the grade beam, the pier dangled by its rebar. The Bobcat operator positioned the bucket under the pier, and I cut the steel with a reciprocating saw. The weight of the pier dropping into the bucket was enough to rock the Bobcat off its rear tires momentarily.

Temporary supports—The door was now open to the crawl space, and McGee could do his work. After carving out the earth between the grade beams, I had him scrape away the soil be-

The dirty work. **Their bases embedded in concrete, 3-in. steel columns carry the weight of the house as the floor of the basement is readied for a concrete slab. In the background, three concrete meter boxes begin the stack that will house the sump pump.**

tween each pier for a temporary 6x6 post atop a 4x12 block on its side. To spread out the load, I put a 6-ton hydraulic jack on top of each post and a 2-in. square by ½-in. thick steel pad on top of the jack's piston. Then I cranked on the jack's lever until I could hear the steel digging into the underside of the grade beam. In this fashion I took the load off the piers and removed them one by one with the jackhammer and the saw.

Builders are constantly imagining worst-case scenarios, and I'm no exception. My major nightmare featured groundwater or rain, coupled with an earthquake. After all, this house isn't far from the San Andreas fault. Sure enough, the night after I'd taken out the last pier, it started to rain, and at 3 a.m. we had a modest earthquake. I woke up with a start and figured I'd just lost my liability insurance. But the forest of temporary posts stayed put, and nothing budged under the house.

Steel columns—The working drawings for the cellar didn't consider groundwater to be much of a problem. The section showed a vapor barrier under the slab and a layer of hot-mop tar applied to the inside of the block walls. That was it. Because I'd be fielding the callback, I wanted to make sure that *no* water ever got into a wine cellar full of fancy wines and cabinets.

The drawings called for permanent steel columns embedded in concrete to support the grade beams. My strategy for drainage began with pitching the slope of the earth toward one corner of the cellar (top right drawing, p. 97), so each post would be slightly different in height. Once the earth was sloped, I took measurements for each column and had them made out of 3-in. steel pipe with ¼-in. steel flanges welded to their ends (photo above). Then I picked up some ½-in. steel plate at a salvage yard and had it cut into 1 ft. squares. I put the squares on the earthen floor next to each temporary column and placed a hydraulic jack atop each plate (bottom right drawing, p. 97). The columns bear on the jacks, but before I put them in place, I slipped an 18-in. long section of Sonotube over each jack to act as a form to encase each jack in concrete. I figured

the weight of the concrete around each jack would be enough to anchor the columns against any accidental construction bumps.

I carefully plumbed the steel columns and centered them under the grade beams. At this installation stage they were held fast against the bottoms of the beams by the pressure of the jack. To make sure the columns didn't move at their tops I ran a couple of ½-in. masonry anchors through the flanges of the columns into the bottoms of the grade beams. Then I pumped concrete into the Sonotubes, entombing the hydraulic jacks, and removed the temporary posts.

Drainage—Although the drawings didn't call for it, I decided that a sump pump was the insurance I needed against water problems. I readied a place for one by pouring a 30-in. square, 1-ft. thick reinforced concrete slab at the low corner of my excavation. On top of this pad, I stacked 18-in. by 24-in. concrete boxes—the kind used for water meters—just above the top of the excavation.

A ring of 4-in. perforated pipe encircles the excavation, terminating at the sump. Water-meter boxes have thin sidewalls that are designed to be knocked out with a hammer to admit pipes. I took out a couple of them to give groundwater a path into the sump boxes. To keep the gravel backfill from getting inside the sump and potentially clogging the pump, I covered the knock-outs with the kind of plastic drain grills that are used in patios to keep leaves out of drains. I used masonry caulk to attach them to the boxes. At the bottom of the sump, I placed a 100-gpm pump with a mercury switch. It starts pumping when the water level rises to 1 ft. above the pump pad. The pump feeds a 2-in. drain line that runs out downhill from the house in the backyard. The pump plugs into a GFCI outlet affixed to a floor joist above the sump. The pump is tied to a galvanized chain, and there is plenty of crawl space around the top of the sump if the pump ever needs maintenance or replacement.

Slabs and speed blocks—Once I had the sump installed, it was time to level the floor and bring it up to the correct height for the 12-in. thick slab foundation. A 12-in. thick slab may seem like overkill, but a slab that is loaded on its perimeter will often crack in the middle. Thickening the slab in the center and using rebar instead of wire mesh reduces the chances that the slab will crack. With the indispensable help of a Yanmar model #YN2260 tractor, we hauled 27 tons of pea gravel down the ramp (photo above).

The slab is reinforced with #5 rebars on 16-in. centers in both directions. At the edges the steel extends up 2 ft. for stubs that overlap the vertical rebars in the walls. I put a vinyl waterstop (a strip of vinyl that serves as a water barrier) around the edges of the slab to discourage water from coming through the cold joint between the slab and the block walls.

The payoff. A mahogany stair curves into the new wine cellar, which is lined with redwood wine racks and cultured cobblestones made of stucco.

The walls are 12-in. speed blocks (bottom left drawing, p. 97) filled with concrete. Unlike a standard block, a speed block is open on both ends, and its single web extends only part way up the sides of the block, making a shelf for horizontal steel. Because their ends are open, the speed blocks were easy to thread around the steel columns. To accommodate the waterstop, the bottom course was installed upside down.

The masons filled the block walls two courses at a time with site-mixed concrete. I had them stop the walls a foot below the bottoms of the grade beams, leaving about half of the top course of blocks unfilled. That way I wouldn't end up with a cold joint between the grade beam and the block walls. I also asked the masons to leave out a few blocks in the top two courses in each of the interior walls and a 4-ft. high opening in the exterior wall next to the Bobcat ramp. These openings would be the route taken by the drain-rock backfill.

I topped off the walls with concrete held in place by forms nailed to the masonry with powder-actuated fasteners. The interior forms were simply lengths of ¾-in. plywood affixed to the face of the grade beams and the block walls. The backside forms were a little more tricky because the walls and the grade beams are different thick-

nesses. These forms were held fast by 2x6 strongbacks and 1x braces (top left drawing, p. 97). I pumped concrete into the forms through a 3-in. hose poked into cavities that I chipped into the bottom edges of the grade beams. Stinging the wet concrete with a vibrator inserted through the cavities helped to work the concrete into the irregular nooks and crannies. Still, there were some voids between the grade beams and the new walls. I filled them with expanding concrete.

Backfill and one more slab—I investigated many ways to get the drain-rock backfill into the spaces behind the block walls. There were finished floors overhead, so there was no easy way to deliver it from above. I thought about conveyor belts but rejected them as too expensive and too unwieldy. I finally settled on the basics: burley guys. I called up our local employment office and had them send over seven day-laborers. I gave them dust masks, gloves, 5-gal. plastic buckets (with instructions to fill them only one third of the way) and a lot of apples, sandwiches and candy bars. In eight hours they lugged 40 tons of drain rock from the bottom of the ramp to the outside of the cellar walls by way of the windows that I'd left in the tops of the block walls.

The finished floor of the wine cellar was to be slate, and I felt that because of the substantial loading on the edges of the foundation slab, it might develop cracks in the middle over time. So I poured a 5-in. slab over the top of it. The top slab is separated from the lower slab by 2 in. of sand, which isolates the slabs. If the slabs were bonded, cracks in the bottom slab would end up in the top slab and its slate finish. The top finish slab is isolated from the block walls around the edges with ¾-in. asphalt-impregnated fiberboard.

I patched the holes in the walls with blocks and coated the inside faces of the walls with two layers of Thoroseal with Acryl 60 (Thoro System Products, 7800 N. W. 38th St., Miami, Fla. 33166; 305-597-8100). Thoroseal expands when it gets wet and increases in density to form a watertight barrier. Masonry finishes such as tile or stone veneer can be applied over it with thin-set mortars.

With the structural work behind me, I could finally start on the mahogany staircase, the redwood wine racks and the cabinetry (photo above). But that's another story. □

Scott Publicover is a builder in Saratoga, Calif. Photos by author except where noted.

Gunite Retaining Wall
Sprayed-on concrete does more than just line swimming pools

by Ken Hughes

I am a structural engineer who works in the San Francisco Bay Area, and these days a lot of my clients are building on steep hillside lots. The reason is simple enough—all the good flat lots have been taken. For the designer, hillside lots present the challenge of fashioning a building that takes advantage of the views and a floor plan that works in concert with the terrain. On the other hand, the builder is usually faced with

Ken Hughes is a structural engineer with the firm of Vickerman/Zachary/Miller Engineering and Architecture in Oakland, Calif.

extensive excavation work and an unconventional foundation system. But no matter what type of foundation is eventually constructed, these hillside projects often begin with hefty retaining walls.

The wall discussed in this article holds back the earth above a home built by Servais Construction in the Berkeley hills. This company specializes in building finely crafted houses, both on a contract and a speculation basis. Whenever I get a call from Jim Servais, I know I'd better put on my hiking boots to inspect the lot.

As with most of Servais' projects, I was skepti-

cal when I first saw this site. It was almost too steep to walk. Servais wanted to build a spec house on the property, so it was understood from the outset that we had to approach the project with that in mind. If we couldn't figure out a way to stabilize the earth within budget, we would have to abandon the project.

We began by getting a soils report from Subsurface Consultants of Oakland, Calif. They found the soil to be reasonably stable, with weathered bedrock 4 ft. to 6 ft. below the surface. Given this news, we calculated that some excavation near the center of the site would allow a house to be attractively nestled into the hillside. The vertical cuts into the hill, however, would have to be bolstered by retaining walls.

Retaining-wall design—With any retaining-wall design, the objective is to stabilize a vertical cut in the soil as economically as possible, yet achieve a long-lasting structure that satisfies accepted levels of structural safety. For this project, several retaining walls were required. The largest is 50 ft. long and averaged 7 ft. in height with a steep, upward-sloping backfill. This wall is above both the house and the street, about 100 ft. from the nearest driveway.

By looking at test borings from the site, our soils engineer knew that the retaining walls would have to hold back soil made of sandy clay. The wall footings would be in the transition area, where the sandy clay mingles with the weathered bedrock.

Using this information, I designed two retaining walls for Servais. This way he could run a cost analysis for each design, and pick the more economical solution. Both designs used a conventional continuous-spread footing to resist overturning and sliding. Wall A was a cast-in-place concrete wall. Wall B would consist of concrete blocks, reinforced with steel and completely grouted. We considered two other wall types—a concrete crib wall (precast concrete members stacked together like Lincoln Logs) and one made of pressure-treated timbers. We rejected the crib wall because it would have required another subcontractor to build it, and Servais wanted to keep the cost down by building the wall with his crew. Although there is nothing wrong with pressure-treated wood retaining walls, we vetoed the idea because bank loan officers don't always believe in them.

Modifying the design—When the hillside cuts were made, two things became apparent. First, the 7-ft. vertical cut seemed to hold temporarily without sloughing. This was partly because of a long dry spell prior to excavation. Also, the bedrock turned out to be a little closer to the surface than we expected. We also realized that most of our retaining-wall footing would have to be trenched and placed in bedrock. At this point, we reconsidered the wall's foundation.

We decided to discard the conventional footing because of all the pick-and-shovel excavation it would have required into the stubborn bedrock. Instead, we opted for a reinforced-concrete grade beam atop 18-in. dia. piers, spaced 6 ft. apart (drawing and photo at right). A backhoe could have handled the excavation, but since Servais needed a drilling rig to bore holes for the house foundation anyway, it made sense to avoid the expense of one more heavy-equipment subcontractor.

Footing aside, Servais was not looking forward to building this wall. The labor involved in carrying the concrete blocks up the hill by hand would be time-consuming and expensive, and the alternative of casting the wall in place involved transporting, building, placing and stripping a considerable amount of formwork.

After pondering the blocks versus the pour, Servais called and asked, "Why can't we build

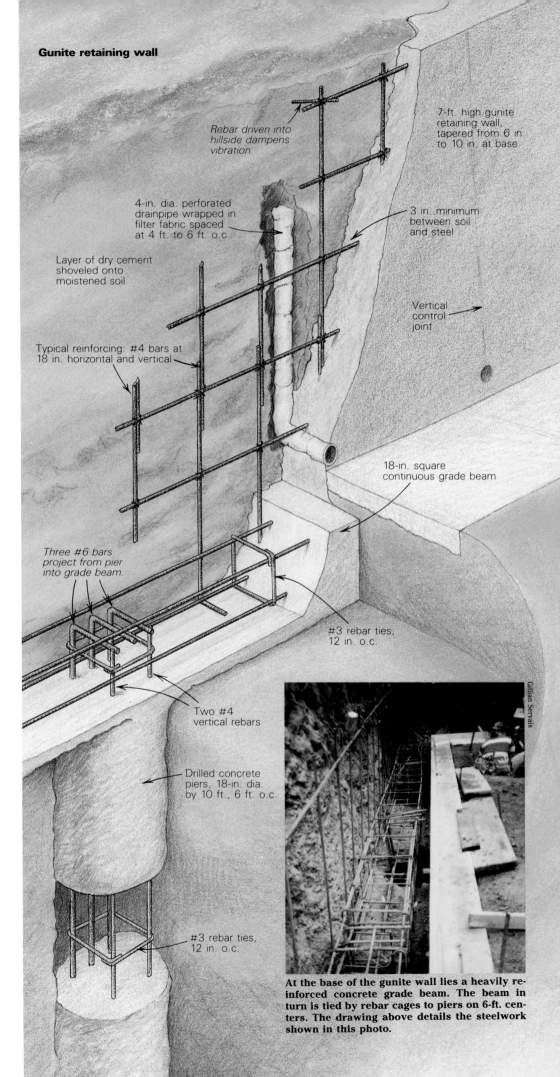

Gunite retaining wall

Rebar driven into hillside dampens vibration.

7-ft. high gunite retaining wall, tapered from 6 in. to 10 in. at base.

4-in. dia. perforated drainpipe wrapped in filter fabric spaced at 4 ft. to 6 ft. o.c.

3 in. minimum between soil and steel

Layer of dry cement shoveled onto moistened soil

Vertical control joint

Typical reinforcing: #4 bars at 18 in. horizontal and vertical

18-in. square continuous grade beam

Three #6 bars project from pier into grade beam.

#3 rebar ties, 12 in. o.c.

Two #4 vertical rebars

Drilled concrete piers, 18-in. dia. by 10 ft., 6 ft. o.c.

#3 rebar ties, 12 in. o.c.

At the base of the gunite wall lies a heavily reinforced concrete grade beam. The beam in turn is tied by rebar cages to piers on 6-ft. centers. The drawing above details the steelwork shown in this photo.

Perforated drainpipes let into vertical cuts in the raw earth relieve hydrostatic pressure behind the wall, left. Before the gunite was applied, they were wrapped with filter fabric to keep them from clogging. The white material on the bare earth is portland cement, which helps to prevent sloughing. At the end of the wall, a minimal form turns the corner. Above, gunite is placed in increments as the operator makes a pass from one end of the wall, then back to the other. The contents of the hose are under tremendous pressure, so it takes a firm, steady grip to control the business end of a gunite hose. If it were to whip about out of control, it could easily injure an unlucky by-stander. A few hours after the gunite crew begins its work, the wall is ready for troweling, below.

this wall out of gunite?'' The idea made sense. It took advantage of the temporary stability of the vertical cut, which eliminated the need for a back form. It would considerably reduce labor and time, compared to the concrete block or conventional cast-in-place concrete. From an engineering viewpoint, gunite is nearly identical to cast-in-place concrete (see the sidebar at right). The steel reinforcing, concrete strength and wall thickness would not change significantly. By making some changes in the drainage system, I was able to adapt the new foundation to work with a gunite wall.

A drainage system behind a cast-in-place concrete or block wall is typically installed after the wall is in place. Not so with a gunite wall. In this case the drainage network is a series of vertical perforated 4-in. dia. pipes let into cuts in the earth on 4-ft. to 6-ft. centers (photo facing page, left). To keep the gunite from clogging the perforated pipes, the crew wrapped them with Mirafi 140N filter fabric (Mirafi Inc., P.O. Box 240967, Charlotte, N. C. 28224). The drainpipes emerge at the bottom of the finished wall, relieving the lateral hydrostatic pressures on it. Before shooting on the gunite, the crew temporarily plugged the pipe outlets with rags tied to rebar handles.

Wall construction—Once the hillside had been cut, it was important to get the wall built right away. This was a winter project, and a heavy rain could seriously have eroded the exposed hillside. To help maintain the bare earth cut, the crew applied a thin layer of portland cement to the vertical surface. They did this by lightly misting the wall with a hose, and then scattering shovel loads of dry cement across the cut. This trick did two things: it helped to prevent erosion of the dirt, and it created a surface to which the gunite would more readily adhere.

Next, the crew tied the steel reinforcing rods in place. The steel is the same size and spacing as for a poured concrete wall, with one exception. Reinforcing steel in a gunite wall has to be secured to keep it from vibrating as the gunite is blasted into position. To dampen the vibration of the steel, Servais' crew tied wires to the tops of some of the vertical rods, and then wrapped the wires around rebar stakes driven into the hill. Without this bracing, the steel might have bounced around as the gunite was blasted onto the wall, causing already placed gunite to fall out in big chunks. Except for a couple of small forms for the wing walls at the corners, the wall was ready to shoot.

The entire 7-ft. height was placed in one four-hour operation (photo facing page, right). As they sprayed the wall, the gunite crew monitered its thickness by watching thin horizontal ''ground wires'' tied to the rebar. As they sprayed the wall, the gunite crew monitored its thickness by watching wire depth gauges tied to the rebar. Typically, the gauges are used as guides for a cutting tool that slices the excess gunite off the face of the wall. This results in a true flush surface. With this wall, however, Servais elected to have a more random finish (bottom photo, facing page) in keeping with the Spanish-style stucco house he planned to build.

As soon as the gunite was in place, the drain-

Gunite and shotcrete

Until 1967, the word ''gunite'' was a proprietary trademark. It is now a generic term used to define the dry-mix shotcrete process. In this procedure, a dry mixture of cement and fine aggregate is pumped through one hose and water through a second. Mixing occurs at a common nozzle where the gunite is ejected at high velocity onto a surface.

''Shotcrete'' is a generic term used to describe the pneumatic placement of any concrete through a hose and nozzle at high velocity. While the term properly covers both the wet-mix and dry-mix processes, the word is used most often to describe the pneumatic placement of concrete in a plastic state.

Recent improvements in the pumps that deliver shotcrete to its target have made it the choice over gunite in some circumstances. While a gunite crew can typically move about 30 cu. yd. of material in a day, a comparable shotcrete outfit can pump about 90 cu. yd. Gunite, however, can be trimmed to a smooth surface, while shotcrete leaves a rough finish that is often plastered for cosmetic purposes.

The birth of gunite—Although gunite and shotcrete came into wide use immediately after World War II, gunite dates back to the turn of the century. In 1895, Dr. Carlton Akely, Curator of the Field Museum of Natural Science in Chicago, developed the original cement gun. He was searching for a method to apply mortar over skeletal frames to form the shapes of full-size prehistoric animals. He could not form the necessary convoluted shapes and contours by conventional troweling, so he developed a method to shoot concrete into place with air as the propellant. In a single-chambered pressure vessel, he placed a mixture of sand and cement. Then he pumped compressed air into the chamber, forcing the mixture into a hose. As the sand and cement mixture was ejected from the end of the hose, it passed through a spray of water that hydrated the mixture.

Immediately following World War II, the use of gunite and shotcrete increased tremendously. Builders found numerous applications in all sizes of projects, from swimming pools to tunnel construction.

Although procedures have been refined and equipment improved, the basic process has not changed since it was originally developed. Gunite or shotcrete can be used in lieu of conventional cast-in-place concrete in most instances, the choice being based upon convenience and cost. These processes are particularly cost-effective where formwork is impractical, or thin layers or variable thickness are required. The principles used in the design of cast-in-place concrete structures are also applicable for gunite and shotcrete structures. Compressive strengths of 2,000 psi to 4,000 psi are common, and higher strengths are easy to attain, depending upon the specific mix design.

Although large civil and industrial projects such as dams, tunnels and aqueducts are the most common use for gunite and shotcrete, other modern applications that are becoming more popular include seismic renovations, basement and shear-wall construction in new buildings, and soil nailing. Old masonry buildings can be strengthened to resist seismic forces by applying reinforcing steel and gunite to the face of the brick, thus forming a strong wall attached to the much weaker masonry. This is usually done on the inside face, which allows the exterior rustic brick facade to remain.

Soil nailing is a relatively new procedure that allows construction of very high retaining walls without the need for a footing or vertical piles. This is a common way to stabilize a deep excavation, such as the perimeter basement walls of underground parking structures below high-rise buildings. This process involves reinforcing the earth by drilling and grouting into place an array of tie-back anchors, typically to a 30-ft. depth. The exposed ends of the rebar strands protruding from the anchors are woven into a reinforced gunite or shotcrete wall that forms the vertical surface of the excavation.

Gunite and shotcrete placement are very specialized operations. The quality of the product is highly dependent on the skill of the workers. The American Concrete Institute (Box 19150, Redford Station, Detroit, Mich. 48219) has prepared and made available ''Guide to Shotcrete'' (ACI-506R-85), which gives detailed guidelines and requirements for successful gunite and shotcrete placements. In addition, the Gunite and Shotcrete Contractors Association (P.O. Box 44077, Sylmar, Calif. 91342) has vast resources of technical data to aid contractors and engineers in the use of gunite and shotcrete. —*K. H.*

---

line plugs were pulled, and vertical control joints were struck into the face of the wall. These control joints project upward from each weep hole. Once the masonry starts curing and shrinking, cracks usually start at the weep holes. They are therefore natural areas to direct crack-control joints.

The last step was to apply curing compound to prevent rapid curing and cracking of the surface concrete. Except for achieving its design strength—in this case 2,500 psi—the wall was complete. Excluding excavation and footing construction, most of the construction was completed in one day, and there were almost no forms to strip or backfilling to do.

Construction costs—This gunite wall was built for just over $2,000, or roughly $6 per sq. ft. of surface area, excluding footing construction. By comparison, a similar concrete-block wall would have cost roughly $7 per sq. ft., plus the costs of transporting the blocks uphill and applying a plaster finish. We estimate that a cast-in-place concrete wall for this project would have cost $9 to $10 per sq. ft. Building and setting the forms would have been labor intensive and costly, and likely as not the form lumber would have been hard to reuse. Perhaps even more important, Servais didn't have to agonize over a tall-wall concrete pour into forms that would have been braced on just the downhill side. □

Selecting a Moisture-proofing System for New Basements

New products from the latest technologies offer more effective alternatives to slopping sticky tar onto new foundations

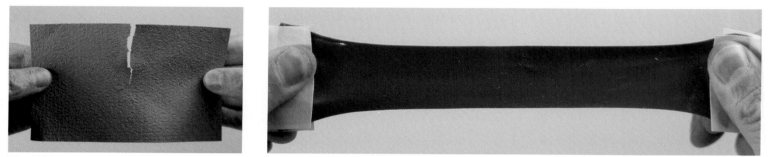

Flexibility means better protection. Flexible cementitious coatings (photo left) might be called waterproof, but their limited elasticity can cause them to crack where foundations do. Elastomeric membranes (photo right), such as RPC's Rub-R-Wall liquid-applied polymer, stretch to span cracks.

by Bruce Greenlaw

Basements rarely conjure up cheery thoughts. They usually remind me of something out of the horror film *A Nightmare on Elm Street*. And when it comes to moisture problems, the truth about basements is a real nightmare. Surveys indicate that between 33% and 60% of the 31,000,000 basements in single-family homes in the United States have moisture problems.

A wet basement is uninhabitable, and the mold or mildew that thrive in such a musty environment can ruin anything stored there. Worse, a wet basement can wick more than 10 lb. to 20 lb. of water vapor a day into a home's interior. In today's airtight homes, this moisture can condense in the building envelope, causing mold, mildew and eventually rot.

Basements don't need to be wet. There are scores of products that, when used appropriately and installed with care, help to keep basements dry. It's important to remember that these products are designed to function in *addition* to proper drainage. (For a detailed look at foundation drainage, see pp. 54-58.)

Waterproofing vs. damp-proofing—Some products are designed to waterproof foundations, while other products only damp-proof them. The American Society for Testing and Materials defines waterproofing as a treatment that *prevents* the passage of water under hydrostatic pressure. Damp-proofing, on the other hand, only *resists* the passage of water in the absence of hydrostatic pressure. Hydrostatic pressure is the force that is exerted on a foundation by the water that is in the ground that surrounds the foundation.

Determining the amount of hydrostatic pressure exerted on any given foundation is essential for choosing the appropriate foundation treatment. One way to predict the water-table fluctuations of a particular building site is to have a soils engineer perform an on-site soil test. A cheaper but probably less dependable way is to ask the neighbors.

Ultimately, the real question is whether it's generally better to damp-proof or to waterproof a basement. The Building Foundation Design Handbook, published by Oak Ridge National Laboratory (for more information on this booklet and other moisture-proofing publications, please see *FHB* #95, p. 136), recommends waterproofing for all habitable basements because most basements are exposed to at least some hydrostatic pressure. But if you're not planning to use your basement as living space, or if the soil around your house is exceptionally well-drained, damp-proofing alone may be sufficient for the basement in your house.

Damp-proofing is less expensive but not as effective—Despite strong evidence that damp-proofing is insufficient in protecting basements from hydrostatic pressure, 95% of all builders damp-proof their basements, according to Micheal Sutton, a regional manager for Koch Materials Company, which makes the Tuff-N-Dri basement waterproofing system. The success rate is questionable: 85% of builders questioned in a National Association of Home Builders survey said that at least some of the basements they have built leak, costing an average of $1,000 to $2,000 per callback.

The standard practice for damp-proofing new basements is to apply one or two coats of unmodified asphalt (asphalt with no chemical additives) to the exterior side of the foundation walls from the footings to slightly above grade. These asphalts come in various grades suitable for brushing, rolling, squeegeeing, spraying or troweling. They cost roughly 40¢ to 60¢ installed per sq. ft. of basement-wall area.

Asphalt emulsions are easiest to work with—Water-based asphalt emulsions can be applied to damp substrates, including green concrete; they aren't flammable; they don't emit noxious fumes; and they clean up with water. They also can be used for gluing extruded polystyrene foam insulation (XEPS) to foundation walls. On the downside, emulsions must be protected from rain and freezing until they have dried, which can take several days in cool weather. Backfilling too soon can cause an uncured emulsified coating to deteriorate.

Cutback asphalts aren't bothered by weather—Cutback asphalts are solvent-based and normally aren't affected by rain or freezing. But most cannot be applied to green concrete or to wet substrates. Uncured, they're toxic and combustible, and they dissolve foam insulation.

The trouble with unmodified asphalt in general is that it is a byproduct of the oil-refining process, and incremental improvements in refining technology gradually have eliminated asphalt's elasticity. Today's asphalts won't span the cracks that invariably occur in basement walls as concrete shrinks and foundations settle or move in response to various other conditions. Asphalts also tend to embrittle or emulsify with age, exacerbating the problem.

According to Brent Anderson, a consulting engineer in Fridley, Minnesota, if you plan to dampproof with asphalt emulsions or cutback asphalts, it is best to apply it 60 mil to 100 mil, or just more than 1/16 in. thick. Gauges for measuring coating thickness are available from Paul N. Gardner Company Inc. (P.O. Box 10688, Pompano Beach, Fla. 33061; 305-946-9454). Some builders improve the moisture protection of asphalt coatings by adding a layer of 6-mil polyethylene sheeting over the asphalt. Anderson says this treatment is an improvement but only if the top edges of the poly are carefully sealed to the foundation, and if the poly is protected from damage during construction.

Rubber damp-proofing is more elastic—Asphalt isn't the only material used for dampproofing. One alternative is Rubber Polymer Corporation's new Rubber-Tite Damp Proofing Plus (for manufacturers' addresses, see sidebar p. 109), which is the only 100% rubber polymer damp-proofing I know of. This type of dampproofing is much more flexible and better at spanning cracks than unmodified asphalt, but it degrades in sunlight and needs to be covered fast. It costs about 40¢ to 50¢ per sq. ft. installed, and it must be sprayed on by certified appliers.

Cementitious coatings are durable but don't stretch—Polymer-modified cementitious coatings such as Masterseal 550 (Master Builders Inc.) and Thoroseal Foundation Coating (Thoro System Products) also are used for damp-proofing. They cost about 20¢ to 65¢ per sq. ft. for materials, or about $1.25 to $2 per sq. ft. installed. These cementitious coatings occupy the gray area of moisture-proofing. Typically brushed or troweled on, they bond tenaciously to cured substrates and, in some cases, even stand up to hydrostatic pressure. They also breathe, which helps prevent basement condensation; they require no protection board; and they look good where they're exposed.

The trouble is, even the best polymer-modified cementitious coatings don't reliably bridge

Liquid-applied membranes are usually sprayed on. Liquid-applied polymer-modified asphalts such as Koch's Tuff-N-Dri emulsion cure to form seamless, self-flashing waterproof membranes over concrete or masonry. Photo courtesy Koch Materials Company.

shrinkage and settling cracks (left photo, facing page). Instead, they tend to crack where foundations do, admitting water where protection is needed most.

When cementitious coatings are used as part of a built-up system that includes rigid-foam insulation, mesh reinforcement and synthetic stucco coatings, such as Sto's Below Grade System, they might truly waterproof foundations. But regardless of what manufacturers claim, cementitious coatings generally are considered by building codes as damp-proofing, not waterproofing.

Waterproofing keeps all moisture out—If you elect to waterproof instead of damp-proof, there are several factors to consider when choosing an appropriate material. Products can differ in everything from shelf life and compatibility with form-release agents to ease of application and cost. Some membranes require protection against backfill; others don't. Also, some membranes can be installed by the average contractor, while others have to be installed by factory-certified technicians.

Laminated-asphalt basement waterproofing is similar to built-up roofing—Laminated-asphalt waterproofing consists of two or more layers of a hot-applied or cold-applied unmodified asphalt, reinforced with alternating layers of a fiberglass or asphalt-saturated cotton fabric. Hot-applied laminated asphalts are seldom used to-

Peel-and-stick membranes stretch to bridge cracks. Self-adhering rubberized asphalt membranes such as W. R. Grace's Bituthene withstand tremendous hydrostatic pressure and can be backfilled immediately. Joints between footings and walls are covered with mastic or mortar for durability. Photo courtesy W. R. Grace & Company.

Thwarting condensation. An insulating board provides protection from backfill and helps eliminate interior condensation. Photo courtesy Dow Chemical Company.

day because they generate nasty fumes and pose a fire hazard during installation. Some companies, however, still promote cold-laminated asphalts as basement waterproofing. These asphalts are essentially the same ones used in damp-proofing, only they are installed in conjunction with fabrics designed to reinforce and augment moisture-proofing. Henry, for example, sells a fiberglass fabric for laminating its #107 asphalt emulsion.

The system has been approved by the city of Los Angeles for use as below-grade waterproofing, with the number of laminations used based on the amount of hydrostatic pressure.

Laminated asphalts cost less than $1 per sq. ft. for materials for a two-ply membrane. They span cracks better and are more durable than unlaminated asphalts, but they're inferior to many other waterproofing systems on both counts. Hot or cold, installation can be awkward, messy and time-consuming. Given the questionable long-term performance of unmodified asphalts, it might pay to take a hard look at the alternatives.

Liquid-applied elastomeric membranes are the most widely used waterproofing systems—Sprayed-on liquids that cure to form elastic membranes are probably the most popular basement-waterproofing products on the market, and for good reason. Not only can they be applied quickly to concrete or masonry (an experienced person can apply about 1,000 sq. ft. per hour), but they cure to form seamless, self-flashing membranes. These membranes conform to complex surfaces, such as curved walls or walls that have a lot of lines or pipes passing through, and they span cracks up to 1/16 in. wide. Also, because the entire membrane bonds to substrates, leaks are confined to small areas that can be detected and repaired easily.

Liquid-applied elastomers aren't perfect, though. Successful application requires painstaking preparation of substrates, meaning that voids

must be filled with a nonshrinking grout or mastic; fins and lumps in the concrete must be removed; and surfaces must be clean and dry, or the membrane can blister or pinhole. For optimal performance, the thickness of the membrane must be monitored very carefully during application to ensure compliance with manufacturer's specifications (35 mil to 60 mil is typical). Also, some products must be heated and then sprayed on using expensive gear, which is bad news for do-it-yourselfers. In fact, many of these products require factory-certified application, although the products usually are backed by good warranties.

Polymer-modified asphalts span cracks better and last longer—It's amazing what a little rubber can do for asphalt. The addition of rubber polymers dramatically increases asphalt's elasticity and longevity. At the same time, these sprayed-on modified asphalts are less expensive than many industrial-strength alternatives.

The two-part Tuff-N-Dri waterproofing system sold by Koch Materials consists of a polymer-modified asphalt emulsion, plus Warm-N-Dri semirigid fiberglass panels. Installed by certified appliers only, the Tuff-N-Dri membrane is applied with an airless sprayer (photo p. 105) at ambient temperatures down to 20°F. It adheres to green concrete and heals itself if punctured. The Warm-N-Dri panel protects the membrane from backfill, conveys groundwater to perimeter drains to prevent the buildup of hydrostatic pressure and insulates, which on the exterior side of basement walls helps eliminate interior condensation. This system is ideal if you live in a cooler climate and plan to use your basement for living space. The system costs about 90¢ to $1.90 per sq. ft. (including labor), depending on your location and the system's R-value, and it carries a 10-year limited warranty.

Mar-Flex's Mar-Kote Drain & Dry Waterproofing System is similar to Tuff-N-Dri, except that its

modified asphalt contains aromatic solvents that evaporate completely and won't leach into the soil after backfilling. It can be applied at temperatures down to 0°F. This system costs about 90¢ to $1.20 per sq. ft., installation by certified contractors included. For the past seven years, Terra-Dome Corporation in Grain Valley, Missouri, has had excellent results using a hybrid system in the poured-concrete underground homes it builds. The heart of its system is a neoprene-modified asphalt emulsion made by Technical Coatings, called ADF-100. ADF-100 can be brushed, rolled, squeegeed or sprayed. Once cured, it stretches up to 2,000% to bridge cracks. For extra moisture protection, Terra-Dome trowels ADF-500 mastic over cold joints before ADF-100 is applied, then lays strips of a bentonite sheeting called Paraseal over the joint locations (more on bentonite later). The whole thing is covered with an insulating shell of XEPS foam before backfilling.

Paul Bierman-Lytle, a New Canaan, Connecticut, architect and builder whose specialty is nontoxic houses, swears by yet another modified asphalt called Safecoat DynoSeal. Made by American Formulating and Manufacturing, it's a neoprene-modified asphalt emulsion that's designed for use by and for chemically sensitive people and for anyone else concerned with using nontoxic products. It can be applied only at ambient temperatures ranging from 45°F to 90°F, though, and costs more than the other modified asphalts: $1.12 to $1.20 per sq. ft., plus application. Bierman-Lytle has used DynoSeal in conjunction with a top-notch subsurface drainage system for the past six years and said he has had great results.

Liquid-applied polymers have the best stretch—Rubber Polymer Corporation's literature shows a guy trying to punch through a cured sample of the company's Rub-R-Wall waterproofing membrane. Instead of breaking, the green membrane stretches the length of his arm.

Although the dynamics of a right jab and a cracking foundation are dramatically different, it is the stretchability of rubber polymers that is their greatest attribute.

Rub-R-Wall is heated and then sprayed by certified contractors over concrete, masonry or rigid-foam foundation forms to a cured thickness of 40 mil. The membrane can be applied to frozen (but not icy) substrates at ambient temperatures down to 15°F. Drainage board or rigid-foam insulation can be attached to its sticky surface 15 minutes after application, and the surface stays tacky for several days. The bond is tenacious, though, so boards and panels must be positioned right the first time. (I know of one guy who used Rub-R-Wall to glue a detached heel back onto his boot.) Rub-R-Wall isn't UV-stabilized, so it needs to be covered promptly. Backfilling can proceed as quickly as 24 hours after application.

In its liquid state, Rub-R-Wall is flammable and toxic, but it's nontoxic when cured. Once cured, it stretches up to 1,800% to span foundation cracks (right photo, p. 104). Independent tests project a 100-year, below-grade life span for Rub-R-Wall, and RPC backs this up with a lifetime limited warrantee. Installed price ranges from about 90¢ to $1.50 per sq. ft. RPC also makes a low-cost alternative to Rub-R-Wall called Graywall. It stretches as much as 1,400% instead of 1,800% and is half the price of Rub-R-Wall.

For builders who can't backfill right away and where UV deterioration is a concern, Mar-Flex recently introduced a 100% polymer product, similar to Rub-R-Wall but with UV stabilizers, called Sunflex. This product's bright-yellow membrane allows limited exposure to sunlight before backfilling. According to Mar-Flex, the installed cost of this membrane ranges from 70¢ to $1.15 per sq. ft.

If you're unlucky enough to get stuck with waterproofing a basement at -10°F, Poly-Wall might save the day. Made by Poly-Wall International, it's a mineral-fortified thermoplastic polymer that can be sprayed, brushed or rolled over cured or uncured concrete or unparged masonry at temperatures down to -10°F. Poly-Wall cures to a glossy gray color (or it can be painted) so it looks good above grade. It requires no protection board and can be backfilled within four hours during summer or the next day during winter. Applied by certified contractors only, it costs just 45¢ to 60¢ per sq. ft., including labor, and has a life expectancy of more than 50 years. What's the catch? Poly-Wall is only slightly elastic, so it doesn't span cracks as well as most other liquid-applied membranes do.

Sheet-applied elastic membranes withstand tremendous hydrostatic pressure—Sheet-applied elastomers are best sellers on the commercial market. Home builders, however, generally use them for high-end work only. This usage is because these membranes withstand enormous hydrostatic pressure, which is typically not encountered in residential work, and consequently cost more than most of the other products I've mentioned. Prices can range from $1.50 to $2 per sq. ft. installed.

Dimple sheeting provides drainage. This polyethylene sheeting resists moisture while the dimples provide air gaps that channel away any groundwater that might get through. Photo by Brent Anderson.

The most common sheet-applied elastomers are 60 mil thick and consist of a layer of self-adhering rubberized asphalt that's laminated to a waterproof polyethylene film on one side and covered by a protective release sheet on the other. Available in 3-ft. to 4-ft. wide rolls, the peel-and-stick membranes install vertically over primed substrates, adhering fully to localize leaks. Edges are sealed by special mastics.

Peel-and-stick membranes offer two layers of waterproofing to bridge cracks. The factory-controlled thickness helps eliminate the thin spots and holidays (skips) that result from spraying. Membranes don't pinhole, can be applied at temperatures as low as 25°F depending on the formula and can be backfilled immediately to eliminate job-site delays.

On the flipside, peel-and-stick membranes don't conform well to complex surfaces, and they're difficult to flash. In addition, concrete substrates must cure for at least seven days before application and must be dry, thawed and free of voids, fins and other defects that could puncture the membrane. Also, time-consuming detailing is required at the inside corners and at the outside corners.

Once the sheets are applied, they can be difficult to reposition. They must be put on perfectly to avoid wrinkles, which need to be cut and repaired before backfilling. Protection board or matting also is required.

W. R. Grace's Bituthene (left photo, facing page) has been on the market for about 25 years, so it's generally considered to be the standard

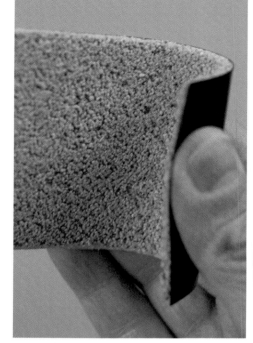

Protection, insulation and drainage material work alongside waterproofing membranes. These boards and mats protect basement moisture-proofing from potentially damaging backfill. The two mats on the right also direct groundwater to perimeter drains, while the middle three boards do that plus provide insulation. Left to right: Amoco Amocor-PB4 fanfolded, extruded polystyrene protection board; Koch Warm-N-Dri semirigid fiberglass insulating drainage panel (sold as a part of the Tuff-N-Dri system only); Dow Thermadry extruded polystyrene insulating drainage panel; GeoTech expanded polystyrene insulating drainage panel; Miradrain 2000R prefabricated drainage composite; and Akzo Enkadrain B drainage mat.

Bentonite clay is mined from the Black Hills. Mameco's Paraseal is made up of granular bentonite clay, laminated to a tough, impermeable high-density polyethylene sheeting. Groundwater causes the bentonite to swell to six times its dry volume, which then forms a waterproof gel.

peel-and-stick waterproofing membrane in the industry. But similar products are made by Karnak, W. R. Meadows, Mirafi, Pecora, Polyguard and Polyken.

The Noble Company sells a different brand of elastic waterproof sheeting. Called NobleSeal, it's a chlorinated polyethylene (CPE) material that comes in 5-ft. wide by 100-ft. long rolls in thicknesses ranging from 20 mil to 40 mil. NobleSeal can be spot-bonded, applied to a grid of adhesive, fully adhered, mechanically fastened or even loosely draped over foundation walls, spanning cracks up to ¼ in. wide (except when it's fully adhered). Seams are chemically welded. One distributor quotes the uninstalled price of NobleSeal at 60¢ to $1.20 per sq. ft., depending on thickness. Installation, however, can double that price.

Bentonite clay is nontoxic and is touted to last forever—Bentonite clay is another relatively expensive waterproofing material that handles great hydrostatic pressure and is used mostly for commercial or specialized residential work. The clay has been used in civil-engineering products since the 1920's and by builders since 1964. It's also used in toothpaste.

The key to bentonite's effectiveness is that it swells up to 15 times its dry volume when wet to form a sticky, impermeable gel in confined spaces. Not only is bentonite supposed to last forever, but it's also seamless, self-healing and nontoxic. It also can be applied in cold weather. Most bentonite products can be installed with ordinary tools over masonry or green concrete

with minimal surface preparation. Foundations can be backfilled immediately, in most cases without protection board, although a drainage layer might be required in some soils. Bentonite also can be an excellent choice for waterproofing slabs in extreme conditions.

On the downside, bentonite must be shielded from rain until backfilling, or it can wash away. Free-flowing groundwater can erode it even after backfilling. Also, some products have limited tolerance to soil salts, alkalis and acids, though salt-resistant bentonite is available.

Bentonite comes in bulk for spray-on application or packed into 4-ft. by 4-ft. cardboard panels that dissolve after backfilling to leave a continuous bentonite membrane. Spray-on bentonite needs special equipment and skill to apply, and I've heard that the cardboard panels can leak at the seams before the cardboard disintegrates. I'd use bentonite sheets or mats instead (photo above right), such as the Mameco Paraseal that Terra-Dome uses. It comes in 4-ft. wide rolls and features granular bentonite laminated to a tough waterproof high-density polyethylene (HDPE) film. Another alternative to Paraseal is Mirafi's Miraclay matting, which costs about 60¢ to 80¢ per sq. ft. for the material or up to about $2 per sq. ft. installed.

San Francisco Bay-area builder Richard Kjelland, who has moisture-proofed foundations with everything from unmodified asphalt to Bituthene, tells me that bentonite sheets come in handy in tight spaces because they can simply be draped down foundation walls. This flexibility makes them a viable alternative where roomy

foundation trenches are impractical. Paraseal can also be used extensively for waterproofing the outside of old rubble foundations, which are too rough and dirty to accept most other types of waterproofing membranes.

Dimple sheeting diverts water while moisture-proofing—Relatively new to the United States but big in Europe, dimple sheeting is a low-cost waterproof membrane that doubles as a drainage mat. Delta-MS sheeting, which is made in Germany and is sold by Intercontinental Construction & Equipment, is a 24-mil, dimpled HDPE sheeting that looks like an egg carton in profile. Available in rolls up to 8 ft. wide, it's simply rolled over concrete, masonry or wood foundations and tacked up with special washered nails (photo p. 107). It can be installed over substrates in any condition and backfilled whenever you're ready. The membrane not only repels water but also forms air gaps against the basement wall that will channel to footing drains any groundwater that might get through the membrane. These air spaces also allow the escape of indoor water vapor that condenses on the outside of the foundation. Depending on the quantity you order, Delta-MS costs 25¢ to 40¢ per sq. ft., and installation is supposed to take about one man-hour per 500 sq. ft. The Norwegian-made System Platon, sold by Big "O" Inc. in Canada, is almost identical to Delta-MS.

Additional products augment moisture-proofing—As I've said, some moisture-proofing membranes need to be protected from backfill.

Moisture-proofing manufacturers

Here's a list of the manufacturers mentioned in this article. For a more comprehensive list, see Aberdeen's annual Concrete SourceBook (available for $30 plus shipping from The Aberdeen Group, 426 S. Westgate, Addison, Ill. 60101; 800-323-3550). It's a 688-page directory that not only covers moisture-proofing, but also just about everything else concerning concrete.

For names of local waterproofers, check the Yellow Pages, ask manufacturers or call the Sealant, Waterproofing & Restoration Institute (3101 Broadway, Suite 585, Kansas City, Mo. 64111; 816-561-8230). The SWRI will also field waterproofing questions. If it can't answer a question, it will try to locate someone who can.
—B. G.

AKZO Nobel Geosynthetics Co., P. O. Box 7249, Asheville, N. C. 28802; (704) 665-5050.

American Formulating and Manufacturing, 350 W. Ash St., Suite 700, San Diego, Calif. 92101; (619) 239-0321.

American Wick Drain Co., 301 Warehouse Drive, Matthews, N. C. 28105; (800) 242–9425.

Amoco Foam Products Co., 375 Northridge Road, Suite 600, Atlanta, Ga. 30350; (800) 241-4402.

Big "O" Inc., 254 Thames Road East, Exeter, Ontario, Canada N0M 1S3; (519) 235-0870.

The Dow Chemical Co., Customer Information Center, 690 Building, Door 1, Midland, Mich. 48640; (800) 232-2436.

GeoTech Systems Corp., 22377 Cedar Green Road, Sterling, Va. 20166; (703) 450-2366.

W. R. Grace & Co., 62 Whittemore Ave., Cambridge, Mass. 02140; (617) 876-1400.

Henry Co., 2911 Slauson Ave., Huntington Park, Calif. 90255; (213) 583-5000.

Intercontinental Construction & Equipment Inc. (ICE), 7666 Highway 65 N. E., Fridley, Minn. 55432; (612) 784-8406.

Karnak Corp., 330 Central Ave., Clark, N. J. 07066; (800) 526-4236.

Koch Materials Co., 800 Irving Wick Drive, P. O. Box 2155, Heath, Ohio 43056; (800) 379-2768.

Linq Industrial Fabrics Inc., 2550 W. Fifth North St., Summerville, S. C. 29483; (800) 543-9966.

Mameco International Inc., 4475 E. 175th St., Cleveland, Ohio 44128; (800) 321-6412.

Mar-Flex Systems Inc., 6866 Chrisman Lane, Middletown, Ohio 45042; (800) 498-1411.

Master Builders Inc., 23700 Chagrin Blvd., Beachwood, Ohio 44122; (800) 227-3350.

W. R. Meadows Inc., P. O. Box 2284, York, Pa. 17405; (800) 342-5976.

Mirafi, Division of Nicolon Corp., 3500 Parkway Lane, Suite 500, Norcross, Ga. 30092; (800) 234-0484.

The Noble Co., 614 Monroe St., Grand Haven, Mich. 49417; (800) 878-5788.

Pecora Corp., 165 Wambold Road, Harleysville, Pa. 19438; (800) 523-6688.

Polyguard Products Inc., P. O. Box 755, Ennis, Texas 75120-0755; (800) 541-4994.

Polyken Technologies, 15 Hampshire St., Building 2, Mansfield, Mass. 02048; (800) 248-7659.

Poly-Wall International Inc., 1879 Buerkle Road, White Bear Lake, Minn. 55110; (800) 846-3020.

Retro Technologies Inc., 3865 Hoepker Road, Madison, Wis. 53704; (608) 849-9000.

Rubber Polymer Corp. (RPC), 1135 West Portage Trail Extension, Akron, Ohio 44333; (800) 860-7721.

Sto Industries, P. O. Box 44609, Atlanta, Ga. 30336-5609; (800) 221-2397.

Technical Coatings Division, S. A. D. I., 5536 Business Park, San Antonio, Texas 78218; (210) 666-2777.

Thoro System Products, 8570 Phillips Highway, Suite 100, Jacksonville, Fla. 32257; (800) 327-1570.

U. C. Industries Inc., 3 Century Drive, Parsippany, N. J. 07054; (201) 267-1605.

Fiberboard and even roofing felt are traditional choices, but a wealth of innovative substitutes now gives membranes more than just protection (left photo, facing page), offering insulation and providing drainage.

For backfill protection only, Amoco, Dow Chemical and U. C. Industries sell fanfolded XEPS panels that open into convenient, 4-ft. wide by 50-ft. long blankets. This foam blanket sticks beautifully to tacky, liquid-applied waterproof membranes such as Rub-R-Wall, or it can be glued with compatible adhesives to other properly cured membranes. Fanfolded protection boards are a measly ¼ in. or ⅜ in. thick, however, and therefore they can offer only a limited amount of insulation.

If it's an insulating-protection board you're after, all three companies also sell XEPS below-grade insulation panels. Amoco's panels are available as much as 3 in. thick, while Dow's (right photo, p. 106) and U. C. Industries' come as much as 4 in. thick. These panels need their own protection where they extend above grade to prevent UV and baseball degradation. Retro Technologies makes a protective, stuccolike coating that can be brushed, rolled or sprayed over the exposed foam.

Besides Koch's Warm-N-Dri and Mar-Flex's Drain & Dry, other products are available that protect, insulate and provide drainage. Dow Chemical's Thermadry is a 2-ft. by 8-ft. tongue-and-groove XEPS panel with a grid of channels on one side that direct groundwater to perimeter drains without affecting the panel's R-value. The channels are covered by a spin-bonded filter fabric that admits water but keeps soil out. Panel thickness ranges from 1 in. to 2¼ in., with R-values ranging from R-4.4 to R-10.6. U. C. Industries' similar Foamular Insul-Drain board comes in 4-ft. wide panels.

GeoTech's Insulated Drainage Board/Panel, on the other hand, consists of expanded polystyrene beads bonded into a 4-ft. by 4-ft. panel that is available in thicknesses up to 2 ft. thick, although 2 in. is usually adequate for most residential applications. Groundwater drains between the beads. The panels are available with or without a filter-fabric skin.

Geocomposites are other products that protect foundations from backfill and hydrostatic pressure, but they don't insulate them. Most geocomposites don't look very sophisticated, but they've earned their stripes in tough commercial applications. The most common geocomposites consist of an impermeable dimpled plastic sheet with a filter fabric glued onto the dimples. The sheets measure ¼ in. to ¾ in. thick, come in 2-ft. to 4-ft. wide rolls and are either nailed, glued or taped up with the dimples and filter fabric facing away from the substrate. After backfilling, the dimples and fabric team up to form silt-free sub-surface waterways. Compressive strengths range from 10,000 lb. to 15,000 lb. per sq. ft., strong enough to restrain backfill without collapsing. Manufacturers I'm familiar with include American Wick Drain, Linq Industrial Fabrics, W. R. Grace and Mirafi. Price varies, but Mirafi's residential Miradrain 2000R geocomposite costs about 35¢ to 45¢ per sq. ft.

Akzo's Enkadrain is a different type of geocomposite that's quickly gaining ground in the residential market. The residential version, called Enkadrain B, is a 0.4-in. thick mat composed of a systematic tangle of HDPE filaments bonded on one side to a filter fabric. It comes in 39-in. wide by 100-ft. long rolls and, where I live, costs about 30¢ to 35¢ per sq. ft. □

Bruce Greenlaw is a contributing editor of Fine Homebuilding. _Photos by the author unless otherwise noted._

Small-Job Concrete

Site-mixed mud can be batched as accurately as ready-mix, given a strong back and a few guidelines

by Bob Syvanen

Most batch plants charge extra for less than a cubic yard or two of concrete. The service you're likely to get on a small order is pretty minimal, so it often makes sense to mix on site. You can do this by hand in a trough or wheelbarrow, or you can rent a mechanical concrete mixer. You'll be using the same ingredients as the batch plant; if you measure and mix carefully, the quality of the concrete should be at least as good. For folks beyond the range of ready-mix trucks, this is the only way.

Ingredients—Clean water is a must for concrete. Sea water is okay if the concrete won't be reinforced with steel. The pour will attain high early strength, but will not be as strong in the long run. Increase the cement content and reduce the water to recover some of the strength lost to the salt.

Cement should be bought by the 94-lb. bag and kept dry. In most cases, Type I is what you need. For resistance to freeze-thaw cycles in northern climates, buy air-entrained cement, which requires using a portable power mixer, since the air-entraining agent needs vigorous mechanical agitation to be effective.

Large aggregate and sand can be purchased by the cubic yard from quarries, building-supply yards and batch plants. Large aggregate can range from ½ in. to 1½ in. It should be clean, hard, durable gravel or crushed stone such as granite or hard limestone. Most sandstone isn't usable. Crushed stone should be square, triangular or rectangular in shape. Flat, elongated pieces shouldn't be used. Sand, the fine aggregate, should be a mix of coarse and fine grains up to ¼ in.

When ordering cement and aggregates, keep in mind that the amount of concrete you get from mixing the ingredients is not nearly as much as the sum of the volume of those materials. This is because the sand in the mix fills in the voids between the gravel or stone, and the cement nestles in between the particles of sand. A rule of thumb that's sometimes used is to figure the amount of your mix to be slightly greater than the amount of coarse aggregate you're using. The chart below will get you a little closer, so you don't end up short on the last batch of your pour.

Unless you are pouring just a few cubic feet of concrete, which can be done using dry-packaged pre-mix, have the aggregates and cement delivered. These materials are extremely heavy, and your good old pickup truck can easily get overloaded with a half-yard of wet sand. Most suppliers will deliver with a dump truck that can spot the materials almost anyplace. As near as possible to where you'll be mixing and pouring is best. Lay plastic sheeting, 6 mil or heavier, on the ground for each kind of aggregate. Plywood is even better because it is a harder surface for scooping against with a shovel, but don't plan on using the plywood for anything very important afterward. You can also use old plywood to separate the sand and coarse aggregate piles vertically, so that they can be placed close together without mingling. Stack bags of cement close together, and up off the ground so they don't turn to stone. Cover them with waterproof plastic whether or not it looks like rain. Don't use cement that is so hard it won't crumble in your hand.

It's okay to scavenge aggregates for your concrete as long as you test them to be sure that they are clean and free of fine dust, loam, clay and vegetable matter. The beach is a good place to find clean sand, and old quarries and stream beds often have acceptable gravels. Aggregates taken from tidal areas contain larger quantities of salt, and should be washed with fresh water before being used in concrete. You can test both sand and gravel for dirt or loam by placing them in a glass jar filled with water. Put on the lid, shake the jar, and then wait for the water to clear. If silt covers the gravel or sand, it needs washing.

There are two tests for vegetable matter. For gravel, add a teaspoon of household lye to a cup of water in a glass jar, add the gravel and shake well. If the water turns dark brown, the gravel needs washing. This can be done with a good hosing. Sand is tested by putting it in a clear glass jar with a 3% solution of caustic soda, which can be made by dissolving 1 oz. of sodium hydroxide in a quart of water. If the solution in the jar remains colorless, the sand is in good shape. A straw color is still okay, but anything that resembles brown means finding another source for sand.

Sand shouldn't be rejected because it's holding a lot of water, but you need to know how wet it is in order to adjust the water content of your mix. Although damp sand feels a little wet, it won't leave much moisture on

| Cement | + | Sand | + | Gravel | = | Concrete |
bag		cu. ft.		cu. ft.		cu. ft.
1	+	1.5	+	3	=	3.5
1	+	2	+	3	=	3.9
1	+	2	+	4	=	4.5
1	+	2.5	+	5	=	5.4
1	+	3	+	5	=	5.8

your hands, and won't form a ball when squeezed in your fist. It contains about ¼ gal. of water per cu. ft. Wet sand will form a ball, but still won't leave your hands very wet. Most sand falls into this category. It contains about ½ gal. of water in each cu. ft. Very wet sand is obviously dripping wet and holds about ¾ gal. of water per cu. ft.

The mix—The strength and durability of the concrete that comes out of your wheelbarrow or mixer depends on the proportion of the cement to the aggregates, and on the proportion of water to cement. Instruction manuals and construction textbooks often show concrete mixes as a ratio of cement, sand and gravel by volume, such as 1-2-4. The first number always represents the cement content, the second is the small aggregate (sand), and the third is the large aggregate (gravel or rock). The more cement used, the stronger the mix. A rich mix, 1-1½-3, is used for roadbeds and waterproof structures. The 1-2-4 mix is used for industrial floors, roofs and columns. A medium mix (1-2½-5) is used for foundations, walls and piers. A lean mix such as a 1-3-6 is used in less demanding applications.

Volume formulas like the ones above give the proportions of dry ingredients but leave the water content up to you. Start with a trial batch, and use the least water you can to get a workable mix. Add a little at a time, and keep track of how much you used.

I favor mix formulas that specify the water/cement ratio, which is called a paste. A 5-gal. paste contains five gallons of water for every bag of cement. This includes the water contained in the sand. The lower the water figure in relation to the cement, the stronger and more durable the concrete. Sidewalks, driveways and floors require a 5-gal. paste for durability. A 6-gal. paste is good for moderate wear and weathering such as foundations and walls. Where there is no wear, weather exposure or water pressure to deal with, a 7-gal. paste will do. Footings are typically poured with 7-gal. paste concrete.

Listed in the chart above are the formulas I use for 5-gal., 6-gal. and 7-gal. pastes, including volume amounts of aggregates for each mix. Each mix differs from the others not only in the ratio of water to cement, but also in the amount and size of aggregates. These adjustments are compromises between economy, strength, durability, workability and slump (stiffness of the mixture). The engineered mixes batch plants use for ready-mix concrete make the same kind of adjustments. The first formula for each mix lists the ingredients used with a single, 1-cu. ft. bag of cement. The second formula gives the correct amount of each material for mixing one cubic yard of concrete, which is useful for figuring and ordering cement and aggregates.

The amounts of water, cement and aggregates in these formulas are given by volume—gallons and cubic feet. There are other formulas that give proportions by weight, but I don't like them as well because ultimately you are trying to fill up a given space—the forms—

Five-gallon paste
1 bag cement, 4½ gal. water, 1 cu. ft. sand, 1¾ cu. ft. gravel (⅜-in. maximum);
for 1 cu. yd. of mix: 10 bags cement, 10 cu. ft. sand, 17 cu. ft. gravel.

Six-gallon paste
1 bag cement, 5 gal. water, 2¼ cu. ft. sand, 3 cu. ft. gravel (¾-in. maximum);
for 1 cu. yd. of mix: 6¼ bags cement, 14 cu. ft. sand, 19 cu. ft. gravel.

Seven-gallon paste
1 bag cement, 5½ gal. water, 2¾ cu. ft. sand, 4 cu. ft. gravel (1½-in. maximum);
for 1 cu. yd. of mix: 5 bags cement, 14 cu. ft. sand, 20 cu. ft. gravel.

with concrete. The formulas above also assume wet sand, with its ½ gal. of water per cu. ft. If you use damp sand, increase the amount of water you add by a quart per cu. ft. of sand. Decrease the water content by a quart for very wet sand. These proportions will yield a fairly stiff mix, depending on the size and shape of your aggregate. But you may need to make adjustments, so mix up a small trial batch first. If the concrete is too soupy, you can correct it by adding aggregates. Don't play with the cement or water content. Instead, add 2½ parts sand with 3 parts gravel in small amounts until the mud stiffens up. For the next batch, be sure to deduct the moisture carried by the extra sand from the total water to be added to the mix. If the test batch is too stiff, use slightly less sand and gravel in the next batch.

Accurate measure—The care that you take in proportioning the mix has everything to do with the quality of your concrete. If you are following a formula for mixing that is given by weight, you will need a bathroom scale for careful weighing. For measuring volume in cubic feet., make a 12-in. by 12-in. by 12-in. frame with no handles or bottom. Place it on a flat surface, fill it level, and lift. Cement is easy to deal with because it comes in 1-cu. ft. bags. You can also make a level mark on the side of your wheelbarrow to indicate the 1 or 2-cu. ft. level. In the case of a ratio mix like 1-2-4, use any convenient measure—a shovel, bucket, or box—but don't let your mind wander when you're counting. For water, mark a large bucket for half-gallons and gallons.

Mixing by hand—A lot of concrete has been mixed by hand, but it is a long, backbreaking job worth avoiding for anything more than a few cubic feet of mud. You can mix in a deep (4 or 5-cu. ft.) wheelbarrow or buy a steel or plastic mortar box (about $70) that holds 6 to 9 cu. ft., or you can make your own mixing tray. A large, shallow plywood box lined with metal so that water won't leak away works pretty well. A flat platform works even better because there are no corners for the shovel or hoe to hang up on. However, mixing must be done carefully to avoid losing water on this flat surface.

First, load the tray with a measured amount of sand. Spread the correct amount of cement evenly over the sand and mix them together with long push and pull strokes with a hoe or shovel. Work the large aggregate into this mix with the same method. Make a depression in the center of the mix and slowly add the water. Pull the mix toward the water until the dry material is saturated, and then turn the mud over until it reaches a workable smoothness. Use this method even if you are mixing in a wheelbarrow or box. A mortar hoe is useful if your aggregate is no bigger than ¾ in. It looks like a large steel garden hoe with two holes in the blade, and costs about $15 to $20. A square-point shovel turned over so that the back of it faces away from you works too.

Mixing by machine—Machine mixing is easier than hand mixing, but it's still a lot of hard work. Electric or gasoline-powered mixers with a capacity of ½ cu. ft. to 6 cu. ft. can be rented by the day or week. Electric mixers are the least trouble. If your job site doesn't have power, then rent a gasoline-powered model. If you are going to use a mixer for more than two weeks and you do lots of small jobs involving concrete, consider buying one.

Set up the mixer right next to your sand and gravel, and run a water hose there. If you have chosen a shady spot, both you and your concrete will set less quickly. Load the drum of the mixer with all of the large aggregate and about half of the water in the formula. Start up the mixer and add the sand and cement slowly, along with the remainder of the water. Let the mixer run for about three minutes or until the concrete has become uniformly grey. When you are finished mixing for the day, add a couple of shovels of large aggregate and some water and turn the machine on one more time to scour the inside. Emptying the drum and a final rinse with a hard jet of water will leave the mixer clean.

Cold-weather concrete—Most engineers do not want you to pour concrete at air temperatures lower than 40°F. A lot of good loads have been poured when it's colder than this, but it's a bit of extra work. Both the aggregate and the water can be heated to keep the concrete warm while it's being mixed and poured, but don't heat the cement. If you heat just the water, bring it to a boil in a 55-gal. drum or other container and pour it on the aggregates to warm them. If you are heating the aggregates also, keep the temperature of the water below 175°F, or the cement will flash-set when mixed, and you won't be able to get a finish on the concrete. Aggregates can be heated on a tray of heavy sheet metal. Build a makeshift firebox out of large stones or concrete block underneath the tray, and heat the aggregates separately. Take them off the fire when they are hot to the touch. □

Bob Syvanen is consulting editor to Fine Homebuilding. *Photo by the author.*

Dry-Stack Block

Precision-ground concrete blocks make it easy to build a wall

by Rob Thallon

Designers have been trying for years to develop a mortarless concrete-block system that could be used by unskilled builders. The concrete blocks in use today look quite uniform, but their dimensions actually vary so much that mortar is necessary not just to hold them together, but also to make up for their irregular sizes. Mixing and applying the mortar to the joints in a block wall require skill and time (see article on pp. 46-50), and the process accounts for 20% to 30% of the material and labor in a masonry project. Manufacturers have recently developed mortarless, interlocking block for industrial and commercial buildings. I use it in house construction. It's called dry-stack block, and it can be laid up as easily as the plastic toy blocks in a Lego set.

Dry-stack blocks look very much like ordinary concrete blocks, but they are consistently a full 16 in. long and 8 in. high (regular blocks are an inexact ⅜ in. less in each direction to allow for the mortar joint). During the manufacturing process, the dry-stack blocks I use are sent through a machine that grinds the top and bottom surfaces to a tolerance of 0.005 in. These parallel, exact and smoothly ground surfaces are what allow the block to be laid up so regularly without mortar.

Most dry-stack blocks have interlocking tongues and grooves at their ends to help align and secure them during placement. Besides standard blocks, there are also bond blocks for bond beams (these have knockouts to accept horizontal rebar), and half blocks. Special corner blocks are manufactured without tongues for finished outside corners (drawing, facing page, center). Where the block remains exposed, its edges are usually chamfered to create a hand-tooled corner that's less likely to chip. It is also possible to have the face of the block ground and sealed to create a smooth, marble-like appearance.

There are three essential differences between the ordinary mortar-laid block and the dry-stack. First, the dry-stack method uses mortar only at the joint between the footing and the first course of block. This mortar joint at the base lets you set the first course absolutely level. Second, ordinary block is usually grouted (filled with concrete when the wall is complete) in only the cells containing reinforcing steel (rebar), while dry-stack blocks are usually grouted in every cell. This locks the blocks in place, and also fills the bond beams completely without having to pour them individually (drawing, facing page, top).

Third, you have to be careful with ordinary block walls to be sure that fallen mortar (as distinct from grout) doesn't hang up in the rebar or clog the bond-beam channels. This usually means that you have to build the wall in 4-ft. vertical increments so that the grout completely fills the appropriate cells. A dry-stack wall, however, can be grouted all at one time because there is no mortar to clog the steel or to plug up the cavities. Grouting tall dry-stack walls all at once can save a lot of time, especially if you use a concrete pumper.

When the dry-stack system was first introduced in the Eugene, Ore., area, about half the projects were questioned by the building department. The building official wanted to see calculations proving that the dry-stack system is as strong as a regular block-and-mortar wall. This is reasonably easy to demonstrate by showing that the compressive strength of the block is greater than that of mortar.

Residential applications—I had seen dry-stack block used successfully on several houses before I had the opportunity to try the system myself. I had designed a house for a steep site, with a complex foundation and several retaining walls. It looked as though using dry-stack blocks would allow a significant saving on labor. In addition, my client wanted a warm-colored block, and not having to use

Reaching as high as 12 ft., the finished block-work is ready for the carpenters. A quarry-tile feature strip is visible just below the top course at locations where the walls will act as foundation for the house. The brownish-red blocks used above grade are special order.

mortar meant we wouldn't have to mix colored mortar to match the block.

Before ordering the block, I asked the supplier about various coloring agents, but everything they showed me gave the blocks a bland uniform color—they looked phony. As an alternative, the manufacturer (Willamette-Greystone Inc., P.O. Box 7816, Eugene, Ore. 97403) suggested using scoria, a brownish-red volcanic aggregate found in Oregon's Cascade Mountains. This seemed to be just what I wanted, so I ordered a special run of blocks.

When the blocks finally arrived at the building site, I was surprised and disappointed. Instead of the rich, red-brown color I had expected, the blocks were pink. Evidently a slurry of scoria dust and cement had come to the surface as the blocks were extruded and vibrated during the manufacturing process. We eventually remedied the problem by sand-blasting the finished wall.

With the footings poured and blocks on hand, we began building the walls. Our crew consisted of an experienced block mason and two laborers. I worked part time. The mason and I were anxious to see just how easily the dry-stack block could be laid up—he from a professional's point of view, and I from the perspective of a novice. On this job, we used almost 3,000 blocks and finished the foundation walls and three large retaining walls in about two weeks. The mason estimated that it would have taken four weeks using regular block and mortar.

First course—Getting the first course level is the most important part of the whole process. If you don't get it right, you'll be fighting your mistakes for the rest of the job. So the first rule is to have good footings, flat and within ¼ in. of level.

Mark the corners of your building on the footing, just as you would for an ordinary block-and-mortar wall, and check for square. It's a good idea to lay out at least one wall on the footing without mortar to test the blocks for length. We found that our blocks varied enough in length to accumulate a ½-in. error in a 20-ft. run if we didn't pay attention. By laying out the blocks dry, we could see how big a gap we had to leave between blocks to make things come out even.

After setting the corner blocks in mortar, we stretched out the mason's line and got down to laying the first course. We found that the work proceeded more easily than we expected, because the vertical joints don't require any special attention. This is a boon for the inexperienced mason. All you need to do is to lay two tracks of mortar along the footing, set the block on the mortar and level in both directions (drawing, right). The smooth surface of the blocks makes leveling easy. As

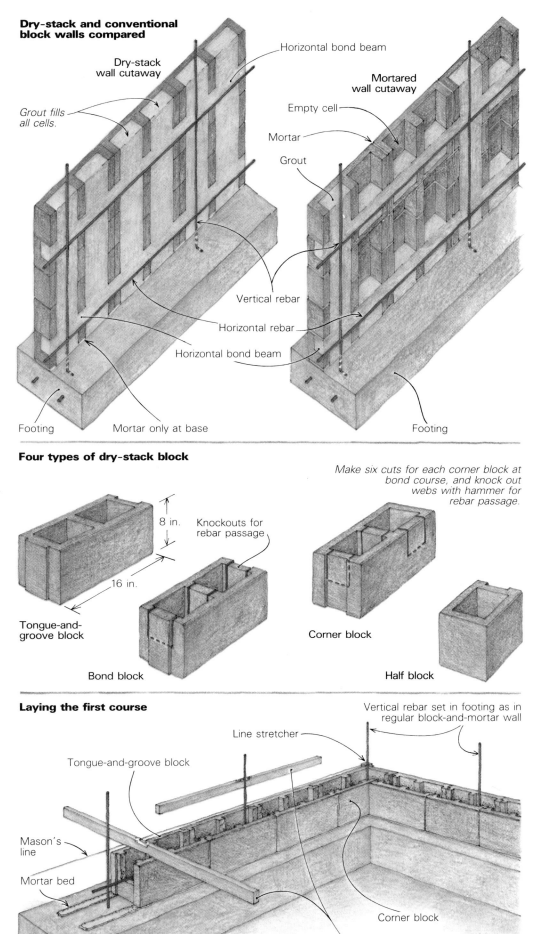

Dry-stack and conventional block walls compared

Dry-stack wall cutaway

Horizontal bond beam

Mortared wall cutaway

Grout fills all cells.

Empty cell

Mortar

Grout

Vertical rebar

Horizontal rebar

Horizontal bond beam

Footing

Mortar only at base

Footing

Four types of dry-stack block

Make six cuts for each corner block at bond course, and knock out webs with hammer for rebar passage.

8 in.

16 in.

Knockouts for rebar passage

Tongue-and-groove block

Bond block

Corner block

Half block

Laying the first course

Vertical rebar set in footing as in regular block-and-mortar wall

Line stretcher

Tongue-and-groove block

Mason's line

Mortar bed

Corner block

Footing as required by soil conditions and code

Level both directions. Differences in level between blocks would cause blocks on next course to rock.

Photos: Rob Thallon; Illustrations: Christopher Clapp

Does it pay? As the chart shows, the dry-stack method costs more for the block itself and needs 30% more grout, but it requires virtually no mortar and saves on labor for grout and laying block. In the project from which the figures were taken, about 10% more was spent for materials, but 24% was saved overall by using dry-stack blocks.

	Dry-stack block walls (actual cost)	Standard block-and-mortar walls (estimated cost)
Materials		
Mortar	$ 30	$ 300
Block	2,847*	2,477
Grout	(32 yd.) 1,481	(20 yd.) 920
Steel (2500 ln. ft.)	461	461
Subtotal	**$4,819**	**$ 4,158**
Labor		
Laying block	$2,700	$ 5,400**
Grout-pump truck	165	(2 lifts) 330
Grout labor	200	300
Subtotal	**$3,065**	**$ 6,030**
Total	**$7,884**	**$10,188**

*2,414 8-in. regular, 345 8-in. half, 175 12-in. regular.   **Based on the mason's estimate of cost-per-square foot at about $2, a conservative figure. The 1979 Western Edition Building Cost File quotes a figure of $3 per square foot.

A quarry-tile inlay makes a thin red stripe around the house. The blocks have been sandblasted and sealed, revealing their volcanic aggregate.

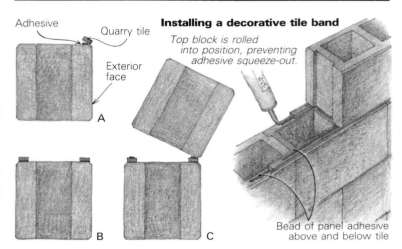

Installing a decorative tile band

Adhesive

Quarry tile

Exterior face

A

B

C

Top block is rolled into position, preventing adhesive squeeze-out.

Bead of panel adhesive above and below tile

we worked, we checked for length every 4 ft. or so and either tightened or loosened the joints slightly to come out even at the corner.

As block walls grow—Once the first course was laid, we built up the corners, carefully plumbed, as a guide for subsequent courses. The weight of the top blocks kept the lower blocks from moving and allowed us to stretch a mason's line tight from corner to corner as a guide. We kept the line about a string's width from the wall, so that accumulating error from successive blocks wouldn't force the string slightly out of line as they touched it. We found that we could set the blocks into place so rapidly that moving the string line became a significant part of the work.

In fact, the work sometimes went so fast that, in our enthusiasm, we made mistakes. The beauty of a mortarless system is the ease with which such mistakes can be corrected. At one point we dismantled a large portion of a 6-ft. tall fireplace footing and ash dump that had gone awry, and put it back together again in less than an hour.

We had some high walls on this project. One, which we built with 12-in. block, was more than 12 ft. tall, and we had several 8-in.

block walls over 8 ft. tall. Walls this high can get pretty wobbly before they are grouted, so we spread a double bead of panel adhesive between every fourth course for stability.

Bond beams—The only major cutting we had to do was notching the corner blocks at bond courses to let in the rebar. We used a mason's saw for this, and for the other minor cutting chores required for vents, bolts and the like. On small jobs, a circular saw with a carborundum blade would work fine.

The rebar required for a dry-stack wall is the same as that required for a block wall with mortar. The minimum requirements are listed in local building codes. For retaining walls up to 4 ft. high or 12 ft. long, we used one #4 bar at 32 in. o.c. vertically and one #4 bar at 24 in. o.c. horizontally (every third course with 8-in. blocks). Beyond these limits we had the wall engineered. Masonry suppliers have brochures listing rebar requirements compiled by the National Concrete Masonry Association.

To form an opening for a window or door, we supported a 2x8 formboard at the appropriate level with temporary posts, then laid up a steel-reinforced and grouted bond beam to serve as a lintel.

A tile feature strip—Just before the last course of block was laid, we installed a narrow tile feature strip. I wanted a thin band of color built into the wall itself to complement the horizontal water-table band at the top of the wall (photo above) so I had a local tile shop split 4x8 red quarry tiles into four lengthwise sections. I sandwiched these between two courses of block. The two pieces with finished edges were used on the exposed side of the wall, and the other two pieces on the hidden side.

We dusted off the top surface of the block and then laid a bead of panel adhesive about 10 ft. long near the inside edge of the tile. I aligned the tiles carefully along the chamfered edge of the block (drawing, above) and rolled them back onto the adhesive, forcing the excess toward the center of the block. This process was then repeated less carefully on the inside of the block and finally on top of the tiles with the final block.

The panel adhesive turned out to be an indispensable part of building the walls. It held all the tiles in position, bonded the top two courses together, and made the wall very rigid. This added rigidity was especially important during the grout pour.

Dry-stack block suppliers

If you're ready to build with dry-stack blocks, you need a local supplier because the blocks are just too heavy to ship economically more than about 150 miles. But finding a local supplier can be frustrating. We called all the block companies listed by the National Concrete Masonry Association as sources for interlocking block and ground-surface block, and asked each company if it could supply the mortarless units. It seems that relatively few builders ask for dry-stacks, so there aren't many made. As a result, the blocks can be hard to find.

The blocks we did find vary in price, design and dimensional tolerances. As a general rule, the closer the tolerances, the more expensive the block. In addition to the basic 8-in. high by 16-in. long by 7⅝-in. wide wall block, each supplier had a full line of sash, corner and bond blocks.

The McIBS Co. is the most active force in the mortarless-block industry. This company has perfected a special liner that can be used in conventional block-molding machines. The blocks made with these liners are double tongue-and-groove on their ends, tops and bottoms, and they maintain dimensions within 0.030 in. This close tolerance, coupled with the tongue-and-groove arrangement on all the hidden surfaces of the block, allows a wall made with them to be sealed with products designed for conventional block walls.

Of the mortarless blocks we found, McIBS are the most widely available, and the most expensive—from 30% to 80% more than conventional blocks. They are currently available in California, Colorado, Illinois, Indiana, Missouri, Nevada, Texas and Wisconsin, with several more states soon to join the list. Write to McIBS Inc., 130 S. Bemiston, St. Louis, Mo. 63105 for more information. For $1, they'll send you a brochure detailing their products and how they are used.

In Texas, builders can find interlocking blocks with tolerances held to ⅛ in. at two places: Valley Builder's Supply, Inc., P.O. Drawer Z, Pharr, Tex. 78577; and the Barrett Co., Rt. 3, Box 211 BI, San Antonio, Tex. 78218. Both suppliers make blocks with tongue-and-groove joints cast into their ends. They cost only about 3% more than conventional blocks. Both companies recommend surface-bonding the finished wall (see the article on pp. 126-129). Similar blocks are available in Minnesota at Charles Friedheim Co., 3601 Park Center Blvd., Minneapolis, Minn. 55416, and in Iowa at the Marquart Block Co. 110 Dunham Place, Waterloo, Iowa 50704.

Oklahoma builders can find dry-stack blocks at the Harter Concrete Products Co., 1628 W. Main St., Oklahoma City, Okla. 73106. Harter grinds its blocks to 1⁄32-in. tolerance, and then cuts a ¾-in. deep slot in the top and bottom of each block to accept a plastic spline. The splines help to align the blocks vertically. The cost is currently about $1.30 per unit for the basic block, and surface-bonding the finished wall is also recommended.

Yet another type of dry-stack block is made by the Buehner Block Co., 2800 S.W. Temple, Salt Lake City, Utah 84115. The block is 8 in. by 16 in., but only 5⅝ in. wide. The blocks hold to ⅛-in. tolerance, and use an interlocking system of plastic rings for alignment during placement.

Dry-stack blocks end up in a wide variety of projects—from houses to roadside sound barriers, from racketball courts to Holiday Inns. They might even become the universal building component that Frank Lloyd Wright and his son Lloyd envisioned in the 1920s. The blocks are simple to use, and they offer the thermal storage capacity that is an essential element of passive-solar design. *—Charles Miller*

Grouting—Before we scheduled delivery, we calculated the amount of grout we needed with the following formulas:

for 6-in. block:
number of full blocks/110 = cu. yd. grout;

for 8-in. block:
number of full blocks /90 = cu. yd. grout;

for 12-in. block:
number of full blocks/50 = cu. yd. grout.

On our job, for example, the calculation was: (2,414 + 345/2 [half blocks])/90 + 175/50 = 32.2 cu. yd. grout.

Because we needed so much grout (about 60 tons), we decided to hire a concrete pumper to get the grout from the trucks to the walls. We completed in about two hours a job that would have taken at least two days if we had done it by hand.

The only problem we encountered was on one of the tall walls. When we filled the cells, the weight of the mud blew out the side of one of the lower blocks, which was probably already cracked. Grout spurted out all over the place. We were able to repair the block and then refill the wall, a little at a time. If you have a wall 8 ft. high or taller, I recommend grouting it up to about 5 ft., filling the shorter walls, and then returning to top off the tall walls after the first pour has had time to set up a bit.

Cleaning—You will inevitably slop some grout over the sides of the block. It's easy to rough-clean the surface by scraping it within 24 hours after the grout is poured. If you have the chance, clean the walls immediately with a light water spray and a soft brush. This can save a lot of hard work later on. To remove the grout stains completely, use a masonry cleaner like muriatic acid.

On this particular job, we wanted to remove both the stains and the pink slurry that formed the surface of our dry-stack blocks. We decided to sandblast only after trying several chemical cleaners without success. The blasting produced the desired results and cost only $320 for the whole job.

Waterproofing—You waterproof dry-stack block the same way as regular block. Below grade, we used Thoroseal, a water-base sealer, which we brushed on in two coats. Some prefer to apply one coat with a trowel, but this requires more skill. I'm sure that any of the asphalt-base sealers recommended by masonry suppliers would also work.

We sprayed the exposed walls with a two-part application of clear acrylic sealer. First, we applied a coat of relatively inexpensive Stone Glamour, then we sprayed on a finish coat of Mex-Seal for the reflective surface we wanted. These sealers bring out the blocks' color much as an oil enhances wood, and protect the block from the deteriorating effects of water penetration. For longest life, an exposed block wall should be resealed every five years or so, depending on the severity of the thaw-and-freeze cycles in your climate.

We didn't do anything to seal the exposed cracks between the blocks, even though we worried about the problems they might cause. I was afraid that capillary action would pull water through the cracks and cause moisture problems inside the house, and that the moisture in the cracks might freeze and fracture the blocks.

We resolved the first problem by sealing the inside of the walls with Thoroseal wherever they enclosed living space. We decided to ignore the second potential problem because the climate here isn't very severe. In the three years since we finished the walls, there has been no cracking.

In a climate where the combination of moisture and freezing is liable to cause problems, I would seal the exposed joints with clear silicone caulk. The caulk could be spread between blocks as the wall is laid up, or applied to the grooves between the blocks' chamfered edges after the wall is assembled. Either one of these procedures would increase construction time and expense, but the job would still be quicker, cheaper and easier than laying up a wall with mortar.

Cost comparison—When we finished the project, all of us who had worked on it were impressed with the dry-stack block. The mason was sure we had cut our labor time significantly by using the dry-stacks (chart, facing page), and he thought that they would result in a 50% labor saving on an average project. The laborers liked it because they got to lay some block themselves, which broke the drudgery of their usual lot—lugging heavy objects around the site all day.

What's wrong with this system?—Availability, that's what. Dry-stack blocks are so heavy that long-distance shipping is prohibitively expensive. Consequently, the blocks have to be manufactured close to their point of use. Although makers of standard concrete block are liberally scattered around the country, relatively few have the grinding or molding equipment necessary for making dry-stacks (see the sidebar above). And unfortunately, there isn't a comprehensive list of manufacturers that make the blocks. So if you're interested in the dry-stack system, get out the Yellow Pages and do some dialing. □

Rob Thallon is a partner in the architectural firm of Thallon and Edrington in Eugene, Ore.

Insulated Masonry Walls

Concrete block comes of age with new technology for energy efficiency

by Bion D. Howard

Masonry construction is a large portion of commercial building, but masonry's share in residential construction has dwindled to less than 10% since World War II, according to the National Association of Home Builders. Only recently have concrete-block walls started to make a comeback in residential construction. This is largely a result of the development of new insulation systems for masonry construction, and of new data that link thermal mass with energy-efficient performance of buildings. Part of my work at the National Concrete Masonry Association (NCMA, Box 781, Herndon, Va. 22070) is to study and evaluate new ways of constructing better insulated masonry walls.

Insulation inserts. Foam insulation inserts like these can be friction-fit into conventional block cores before the block is laid up. They will increase the insulative value of an 8-in. block to at least R-5.

Mass and insulation—Before examining options for insulating masonry walls, it's important to put the current emphasis on high R-values into perspective. The recent interest in superinsulation has focused a lot of attention on resistance insulation, chiefly in the form of fiberglass batts. The notion that larger R-values are the only means of achieving better thermal protection emerged as a fabulous marketing tool for the insulation industry. Unfortunately, this isn't the entire story. The missing parts of the equation are the heat-storage capacity and radiative properties of the walls, which can cause the building envelope to perform differently than steady-state calculations predict. New ways of insulating masonry walls can provide the best of both worlds: reducing conductive heat loss using insulation, and improving heat-retention capacity with mass.

The main types of masonry-wall insulation systems are as follows: lightweight concrete made with insulating aggregates; loose fill insulation that is poured into the concrete-block cores as the wall is built, or "pour-foam" installed in the same way; formed plastic insulation inserts that can be fit into the cores of conventional concrete block, or concrete blocks with specially designed cores; cavity walls where two separate *wythes* (a wythe is a single vertical masonry wall) of masonry are built, creating an air gap that is partially filled with insulation; exterior application of insulation covered with weatherproof coatings after wall construc-

tion; and composite construction units using high-strength adhesive or mechanical bonding of two masonry wythes on either side of a layer of insulation.

These insulation systems for masonry walls can be used alone or in various combinations. Thermal-mass research shows that the greatest energy savings are possible when the insulation is located outside the masonry wall. But this isn't always the best option because of cost or design constraints. Integral insulation systems, which have grown in number since the energy crisis, are an alternative choice and can often be more economical than exterior insulation on masonry walls.

Lightweight, insulated concretes—One method of "pre-insulating" concrete masonry is to change the properties of the concrete itself. Lightweight concrete was invented for shipbuilding during World War I. It is made by using a lightweight aggregate such as expanded polystyrene or glasses made from polymer or mineral bases. The lightweight aggregate decreases the density of the concrete, thereby boosting its insulative value.

As might be expected, some strength is sacrificed with the use of lightweight aggregates. The Sparfil Corporation (5 Veronica St., Box 235, Coburg, Ont., Canada K9A 4K5) spent several years on costly experimental development before it was able to claim that it had a block that was both highly insulated and lightweight. The

Sparfil block relies on expanded polystyrene for insulative value. An 8-in. block is rated at R-6 and weighs just 27 lb. (a standard 8-in. concrete block weighs about 35 lb.). For still greater R-value, these blocks have staggered cores that can be filled with foam inserts. If these are used, a finished wall of 12-in. block can approach R-25. Sparfil walls (top drawing, facing page) are constructed by the surface-bonding technique, using high-strength, fiber-reinforced cement coatings on both sides of the dry-stacked block masonry. Vertical steel reinforcement can also be added to these walls through the cores. The main advantages of mortarless assembly are reduced labor costs, equal or better strength than unreinforced mortared assemblies and very low air-infiltration rates. The surface-bonding compound also creates a sound moisture barrier.

Pour-in core insulation—The principal loose fill materials for block and brick walls are Perlite (Perlite Institute, 6268 W. Jericho Tpk., Commack, N. Y. 11725) and Zonolite (W.R. Grace Corp., Construction Products Division, 62 Whittemore St., Cambridge, Mass. 02140). These are loose beads of expanded polystyrene. Another type of pour-in insulation, called Poly-C (Upjohn Corp., Chemicals and Plastics Research, 555 Alaska Ave., Torrance, Calif. 90503) is poured into the cores as a liquid. A chemical reaction causes it to expand and fill the cores. Frothane (Therma-froth Systems, Inc., 99 Collier St., Suite 300, Binghamton, N. Y. 13901), Kasko K1-10 (Kasko Industries, Inc., 301 West Hills Rd., New Canaan, Conn. 06840), Air-Krete (Air Krete, Inc., Box 380, Weedsport, N. Y. 13166) and Thermal-Krete (Omni-Tech Energy Products, Inc., 1515 Michigan Ave. N.E., Grand Rapids, Mich. 49530-2085) are other brands of insulation that can be "foamed" into the voids in the block cores.

Both Perlite and Zonolite insulations are silicone treated for water repellency, and are fireproof. The plastic-based insulating fills are not fireproof, but are isolated in the masonry cores. You should check local fire codes before installing plastic insulations inside block since there may be special regulations in your area. Today's foam-insulation products claim to be free of urea-formaldehyde emissions, but it's wise to get this in writing from the contractor or supplier, so that a document may be passed on to new owners. Some states require urea-formaldehyde foam-insulation certificates before homes can change hands or be occupied, even though the ban on urea-formaldehyde foam has been lifted.

Formed plastic insulation inserts—Masons, mason contractors and builders seem to like using new concrete blocks with plastic insulation inserts that have been installed at the block plant. Most of these blocks are made from conventional concrete and arrive on site with insulation already in place and ready to lay up. No added labor or insulation is required to reach R-7 to R-10 values.

The Korfil block (photo facing page) comes with a C-shaped insert that is friction-fit into the block cores (Korfil Inc., Box 123, Chicopee, Mass. 01014). Bend Industries (2929 Paradise Dr., West Bend, Wis. 53095) and the Miller Material Co. (Box 1067, Kansas City, Mo. 64141) have developed several similar insert systems (middle drawing, right). Cast and "split" architectural facings are available on some of these insulated blocks.

The Insul Block Corp. (55 Circuit Ave., West Springfield, Mass. 01089) and Formbloc Inc. (Box 546, Concord, N. H. 03301) manufacture cross-web, flat panels of foam that fit into modified concrete blocks. In these systems, the foam insulation extends the full length and nearly the full depth of the block, providing a more complete thermal barrier. In order to accommodate the maximum amount of foam insulation, the blocks are designed with cut-down block cross webs, as shown in the bottom drawing at right.

Korfil has taken the cut-down web method a step further in a new insulated block that should approach R-10. Called NCMA Korfil High-R, the block uses two polystyrene insulating panels per unit. These panels are slightly different sizes and overlap the mortar joint areas to reduce air infiltration and moisture migration. Horizontal reinforcing rods can be used with channels already formed during the block-making process. Full-scale structural testing of High-R walls is now underway at NCMA's research laboratory, and the system should be available on the market in the near future.

Sparfil makes inserts to add to its insulated lightweight blocks. The Sparfil cores are long and narrow, and they're staggered in three rows to reduce thermal bridging. This approach elongates the normally linear flow of heat through the unit, delaying heat loss.

Essenco's E/Block (Essenco Inc., 834 Eagle Dr., Bensenville, Ill. 60106) uses a phenol-based high-R foam that's inserted in special mini-cores in the block (top drawing, p. 119). Only small areas of thermal bridging occur at the block ends. The block is manufactured with a recess at each end that can be filled with site-applied insulation. Essenco claims E/Block also can attain 20% to 30% higher compressive strength than regular block. High ratings on fire, sound and moisture resistance are also claimed. Early in 1985, these blocks were exhibited at the Na-

Drawings: Frances Ashforth

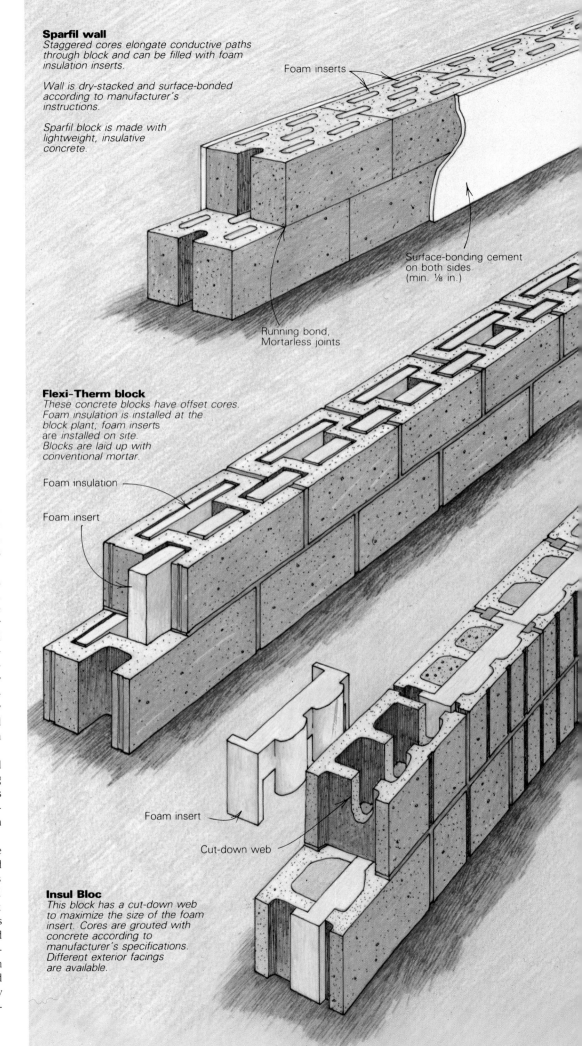

Sparfil wall
Staggered cores elongate conductive paths through block and can be filled with foam insulation inserts.

Wall is dry-stacked and surface-bonded according to manufacturer's instructions.

Sparfil block is made with lightweight, insulative concrete.

Foam inserts

Surface-bonding cement on both sides (min. 1/8 in.)

Running bond, Mortarless joints

Flexi-Therm block
These concrete blocks have offset cores. Foam insulation is installed at the block plant; foam inserts are installed on site. Blocks are laid up with conventional mortar.

Foam insulation

Foam insert

Foam insert

Cut-down web

Insul Bloc
This block has a cut-down web to maximize the size of the foam insert. Cores are grouted with concrete according to manufacturer's specifications. Different exterior facings are available.

Cavity walls. **Two separate masonry wythes, one of brick and the other of concrete block, create an airspace that is filled with rigid insulation. The width of the cavity can vary from ¾ in. to 6 in. or more, depending on how much insulation is desired.**

tional Concrete Masonry Association Concrete Industries Exhibition in Atlanta and met with good reviews by industry experts.

Cavity walls—Cavity walls saw their first use nearly 70 years ago, and we know about them primarily because of more recent demolition work. The brick and block cavity wall (photo above) has several definite advantages over single-wythe masonry. In cavity-wall masonry construction, a continuous air space is left inside the wall between two masonry wythes.

A cavity wall takes more time and material to lay up than a single-wythe wall, but it's the best system in many ways, providing insulation space, thermal mass, sound reduction, fire resistance, and control over moisture and air infiltration.

The cavity wall is built with weep holes in the bottom of the exterior wythe to give moisture an escape route away from the insulation layer and interior masonry wall. Larger screened holes, built into the exterior wythe just under the eaves, can help to vent the wall and can eliminate the need for a vapor retarder.

Cavity-wall construction is good for hot, humid and stormy climates where moisture is a problem and outdoor heat must be repelled. Under these conditions, reflective insulations like foil-faced boards perform well when used in the cavity. If a foil-faced insulation board is used in a cavity wall, the air space must be kept even,

and mortar droppings should be removed from the back of the outer wythe so that they don't touch the insulation.

The air space, or cavity, can be made as wide as 6 in. for low-rise construction, according to recent NCMA structural analyses. This provides enough room for up to 4 in. of foam-board insulation, available in a variety of R-values. A 2-in. air gap (on the exterior side of the cavity) is recommended with a 6-in. cavity. Pour-in insulation can also be used, so the cavity wall is a fairly flexible system. Wall ties 6 in. and longer are now available through construction jobbers, but many current building codes allow a maximum 4-in. cavity, and recommend ¾ in. as the minimum air gap. The cavity-wall builder may need to have a structural engineer perform calculations to support the use of a wider cavity, especially in multi-story construction.

In cooler climates, the vapor barrier in a cavity wall should be located closer to the interior conditioned space. If furring and drywall are going to be used against the exposed inner masonry, polyethylene film can be installed either against the masonry or between the furring and the drywall. Alternatively, the new airtight drywall approach (ADA) could be used, where the interior drywall provides the air barrier. With this system, vapor transmission is reduced by eliminating air leakage through the use of gaskets and careful workmanship.

Composites and composite walls—Composite masonry is typified by rigid insulation sandwiched between brick and block masonry units, with no airspace between the materials (middle drawing, facing page). This arrangement is achieved with high-tech adhesives and mechanical fasteners. It is also possible to use factory-applied expanding polyurethane foams, which are highly adhesive. This system is now in use in England. The advantage of composite wall construction over cavity wall construction is that the wall lays up in one step just as you'd lay up pre-insulated block. A wide variety of exterior appearances can be obtained with less labor than it takes to build a cavity wall.

In most cases, however, composite masonry units are laid up with mortar. In mortared wall systems, the mortar joint will affect the overall R-value of the wall. For example, a wall built with 8-in. by 8-in. by 16-in. composite masonry units rated at R-12 would have an overall R-value of R-8.2 because the R-1.6 mortar joints account for 7% of the wall surface area.

Several composite masonry-wall systems are currently under development. Experiments show it is possible to produce composite units on modified conventional concrete-block machines. Composite masonry units could use low-density insulating concrete, and integral insulation would provide full thermal breaks if the wall were surface-bonded rather than mortared. It is also probably feasible to mix a phase-change material in with the inner side's concrete, enhancing the thermal characteristics of the wall.

Exterior insulation of block walls

Exterior insulation of block walls—This approach can be used for retrofits as well as for new construction (see article on pp. 59-61). The development of exterior insulation systems for concrete block makes it possible to transform aging concrete and "cinder-block" buildings into good-looking, energy-efficient structures. Exterior insulation retrofits can be done with little or no disturbance to those living in the house. The drawing at the bottom of the page shows a typical exterior insulation system.

The typical block wall built before the energy crisis has an R-value of between R-3 and R-7. Adding 1 in. of R-5 exterior insulation and covering this with a protective stucco finish more than doubles the insulative value of the block wall. In addition, the wall mass can better interact with the HVAC system and the indoor environment by storing heat.

Exterior insulation systems (insulation and protective covering) can cost between $3 to $7 per sq. ft. installed. The cost varies with the thickness of the insulation, the size of the building and the particular detailing requirements. You also have to take into account any replacement of windows and doors that should be done as part of the energy upgrade. NCMA's pamphlet TEK 134 describes the exterior insulation of block walls in detail and provides cost data, fuel conversion factors and a map of suitable block exterior insulation levels for the various parts of the U. S. More information is also available from the Exterior Insulation Manufacturers Association (1133 15th St. N.W., Washington, D.C. 20005). EIMA is now setting industry standards, compiling case histories on exterior insulated projects and coordinating industry information.

Cost versus performance—A good way to evaluate insulated masonry wall systems is to compare them with conventional concrete-block construction. Let's assume the unit cost of a conventional, uninsulated 8-in. block wall to be 1.0, and recognize that this wall will be rated at R-0.11 to R-0.29 per in., depending on concrete density. Advanced systems like Sparfil will cost nearly 2.5 units and provide 0.69 to 2.88 R-per-inch, depending on the concrete mix and the insert installation.

Core insert insulations will cost at least 1.4 units and provide an R-per-inch of 0.40 to 0.56. Loose-fill insulations installed on site (Perlite, Zonolite) provide a broader range of R-per-inch, 0.33 to 0.70, at only slightly higher cost.

The cut-down, web-type insulated block appears to be most cost-effective option. Korfil and Insul-Bloc are now producing and/or developing such products. The composite systems, most of which are still experimental, should provide R-per-inch of 0.73 to 3.75 at a cost of 1.7 to 2.0 units. □

Bion Howard is a technical advisor and energy engineer for the National Concrete Masonry Association in Herndon, Va.

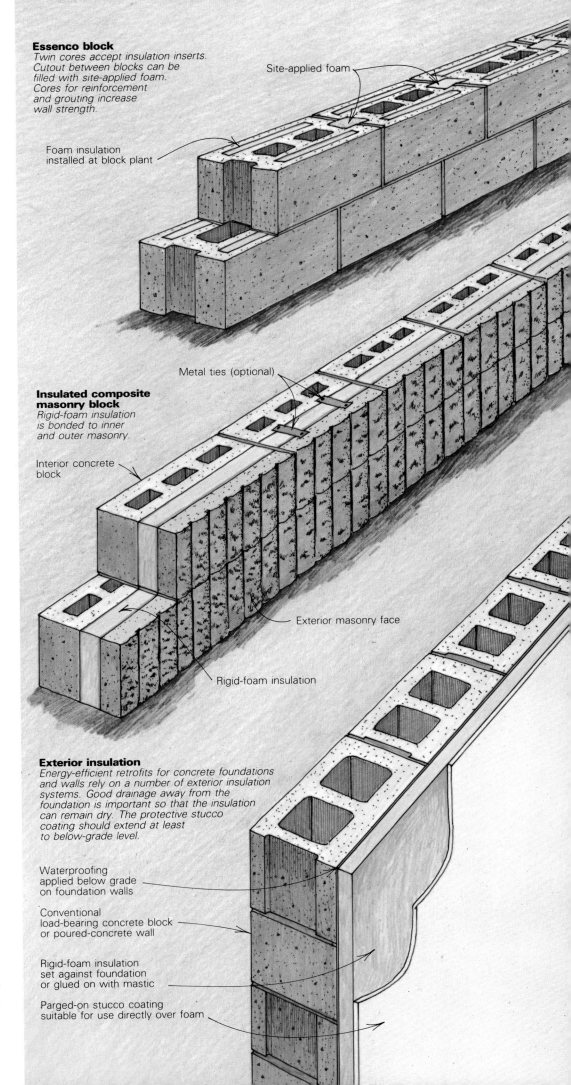

Essenco block
Twin cores accept insulation inserts. Cutout between blocks can be filled with site-applied foam. Cores for reinforcement and grouting increase wall strength.

Site-applied foam

Foam insulation installed at block plant

Metal ties (optional)

Insulated composite masonry block
Rigid-foam insulation is bonded to inner and outer masonry.

Interior concrete block

Exterior masonry face

Rigid-foam insulation

Exterior insulation
Energy-efficient retrofits for concrete foundations and walls rely on a number of exterior insulation systems. Good drainage away from the foundation is important so that the insulation can remain dry. The protective stucco coating should extend at least to below-grade level.

Waterproofing applied below grade on foundation walls

Conventional load-bearing concrete block or poured-concrete wall

Rigid-foam insulation set against foundation or glued on with mastic

Parged-on stucco coating suitable for use directly over foam

Designing Forms for Tall Concrete Walls

Site-built panels that hold up under stress

by William Doran

"**S**top the pour!"

The words echoed through the basement, loud enough to be heard over the transit-mix truck's engine and the grinding pulse of the pumper.

"I've heard that before," said the driver to nobody in particular. "That's the cry of panic."

The problem started when a snap tie snapped at the wrong time. My friend Charles, a builder who specializes in remodeling, was used to building his forms with plywood and snap ties. He'd never had a problem with the system and didn't bother sizing the ties or spacing them to suit different circumstances. He used what he had on hand, and it always held together.

But this time the concrete dispatcher did him a favor by sending three truckloads of concrete in quick succession (he wanted to make up for a year's worth of late deliveries, and it was a slow day at the yard). Filled quickly to their tops, the forms strained against the weight of the soupy concrete. One of the ties popped, and the forms bulged outward. The adjacent ties, already at their limit and now having to pick up the load of the failed tie began to pop in quick succession. In the blink of an eye, the form unzipped, engulfing the laundry room, the water heater and the new furnace in 20 yards of fresh concrete.

Why did the forms fail? Was a tie left out? Were they too far apart? Or was the wall poured too fast? Would stronger ties have made a difference?

In my experience, most carpenters have good instincts when it comes to working with concrete. They know how to build forms and how to brace them against the imposing loads brought to bear by truckloads of fresh transit mix. But I don't know many carpenters who understand the criteria that govern form designs. Few carpenters know what the proper pour rate is or how to proportion the forms to achieve economy in material. Most carpenters overbuild the forms and hope that they'll hang together. Using a form that is significantly overdesigned is almost as bad as pushing an underdesigned form to failure. Both practices waste money, and in the case of failure, somebody might get hurt.

I work for a large general engineering firm in Los Angeles, California, and along with several other engineers, spend a portion of my time designing site-built concrete forms, some of which are very complex. But you don't have to be an engineer to size the forms for straightforward pours—even the 8 ft. to 10 ft. walls that are typi-

cally cast for basements or retaining walls. What is necessary is an understanding of what a form should do, what forces are going to be acting on it, what materials are available for building the form and the limitations of those materials. Once you've got a handle on those variables, you can use the table below to design your own forms.

The drawings on the facing page show the two kinds of site-built forms that I recommend. Both use plywood form panels braced by studs or by horizontal members called walers (usually 2x4s or 2x6s). And both use form ties to hold opposing forms together against the pressure of the fresh concrete. I'll talk about these two systems in greater detail, but before I do, let's look at the materials and the hardware used to build them.

Forming materials—In theory, any plywood designed for exterior use can be used in a concrete-forming application. The smoother the face, the better, especially if you plan to use the plywood as forming material more than once. That's because wet concrete will get into the crevices in the face veneers, and when you pull the panels

away from the wall, some of the wood fibers will stay with the wall. Soon you will have a ragged, weak panel. Interior plywood is not suitable for forming because it delaminates when wet.

A product called Plyform is produced by various mills that are members of the American Plywood Association (APA, P. O. Box 1170, Tacoma, Wash. 98411; 206-565-6600). It is designed specifically for concrete-forming applications. Its veneer is limited to certain species and grades that provide greater uniformity and strength.

Plyform is available in various thicknesses, but we rarely use anything other than ¾ in. unless we are forming curved walls. It costs about 60¢ per sq. ft. Plyform panels are sanded on both sides and have a coating of form oil applied at the mill (form oil is a release agent used to prevent concrete from adhering to forms). If the panels have been idle for some time, however, it's a good idea to re-oil them before use. Also, the APA recommends that the panel edges be sealed before their first use. A book I have on concrete forming says to use aluminum paint, chlorinated rubber-based paint, oleoresinous paint or

DESIGNING STUD-AND-WALER FORMS													
Form thickness	Plywood orien-tation	2x4 stud spacing	Max. concrete pressure	Max. waler spacing	Horiz. tie spacing	Snap tie required	Concrete temp. and rate of placement (ft./hour)						
							40°	50°	60°	70°	80°	90°	
¾ in.	Paral.	6 in.	1,200 psf	29 in.	16 in.	5,000-lb. snap tie	FLH	FLH	FLH	FLH	FLH	FLH	
⅝ in.	Perp.	8 in.	1,200 psf	24 in.	16 in.	4,000-lb. snap tie	FLH	FLH	FLH	FLH	FLH	FLH	
¾ in.	Perp.	9 in.	1,000 psf	24 in.	16 in.	4,000-lb. snap tie	3.8	4.7	5.7	6.6	7.6	FLH	
¾ in.	Paral.	8 in.	900 psf	29 in.	16 in.	4,000-lb. snap tie	3.3	4.2	5.0	5.8	6.7	7.5	
¾ in.	Perp.	12 in.	732 psf	24 in.	24 in.	4,000-lb. snap tie	2.6	3.2	3.9	4.5	5.2	5.8	
⅝ in.	Paral.	8 in.	633 psf	36 in.	16 in.	4,000-lb. snap tie	2.1	2.7	3.2	3.8	4.3	4.8	
⅝ in.	Perp.	12 in.	576 psf	29 in.	24 in.	4,000-lb. snap tie	1.9	2.4	2.8	3.3	3.8	4.3	
¾ in.	Paral.	12 in.	492 psf	33 in.	24 in.	4,000-lb. snap tie	1.5	1.9	2.3	2.7	3.0	3.4	
¾ in.	Perp.	16 in.	412 psf	31 in.	24 in.	3,000-lb. snap tie	1.2	1.5	1.7	2.0	2.3	2.6	

Paral.=Face grain parallel to studs
FLH=Full liquid head (see sidebar p. 52)
Perp.=Face grain perpendicular to studs

This chart shows how you can vary the thickness of the form material, its orientation and the spacing of the studs and the walers to accommodate different concrete temperatures and pour rates. Waler and stud allowable stress is based on #2 Douglas fir, and the forms are a maximum of 8 ft. tall. Walers should be no farther than 1 ft. from the top and the bottom. You can increase the spacing between the top two walers, but no more than 1½ times the normal spacing.

Panelized form systems

Single waler

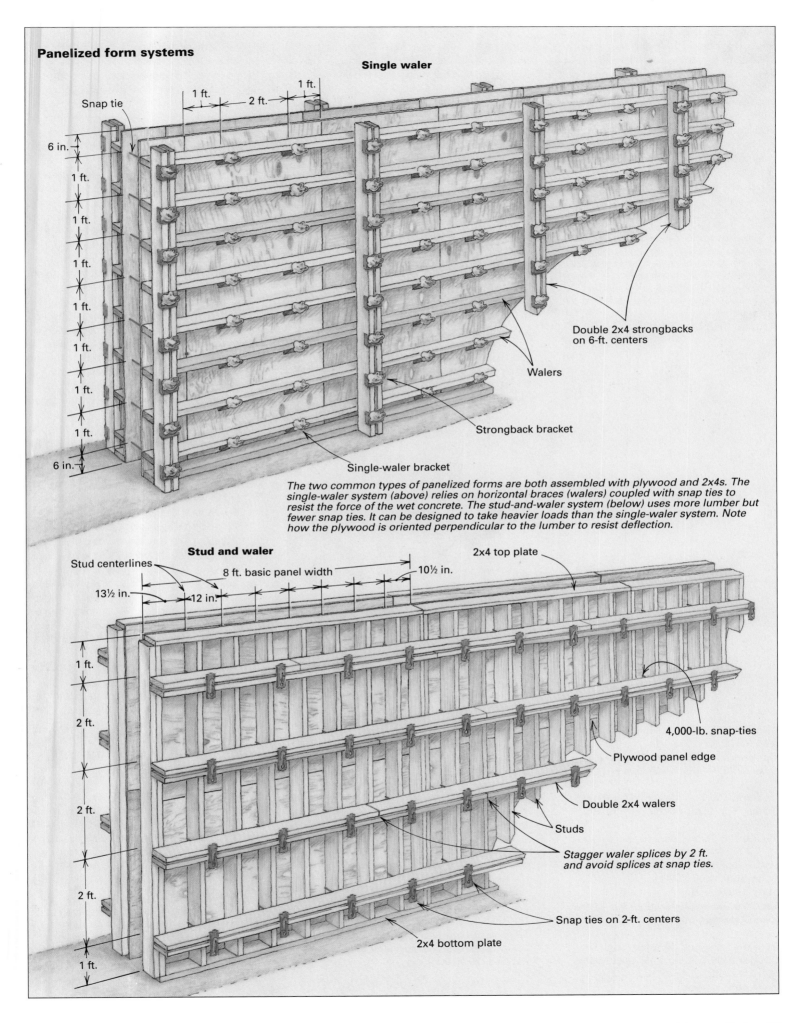

Snap tie

6 in.

1 ft.

1 ft.

1 ft.

1 ft.

1 ft.

1 ft.

6 in.

1 ft.

2 ft.

1 ft.

Double 2x4 strongbacks on 6-ft. centers

Walers

Strongback bracket

Single-waler bracket

The two common types of panelized forms are both assembled with plywood and 2x4s. The single-waler system (above) relies on horizontal braces (walers) coupled with snap ties to resist the force of the wet concrete. The stud-and-waler system (below) uses more lumber but fewer snap ties. It can be designed to take heavier loads than the single-waler system. Note how the plywood is oriented perpendicular to the lumber to resist deflection.

Stud and waler

Stud centerlines

8 ft. basic panel width

10½ in.

13½ in.

12 in.

2x4 top plate

1 ft.

2 ft.

2 ft.

2 ft.

1 ft.

4,000-lb. snap-ties

Plywood panel edge

Double 2x4 walers

Studs

Stagger waler splices by 2 ft. and avoid splices at snap ties.

Snap ties on 2-ft. centers

2x4 bottom plate

urethane coatings for this purpose. But we never seem to have any of those on hand, so we use silicone caulk to seal the edges. It works fine.

According to the APA, it is common to get 5 to 10 uses from a sheet of Plyform, though in my experience the wear and tear of assembling and stripping the forms often take their toll before the face veneer wears out. For purposes of comparing costs with other form materials or rental form systems, I would assume four uses.

Plyform is also available with a special overlay of resin-impregnated material. It's called High Density Overlay Plyform, or HDO. This overlay makes the panel more resistant to moisture and abrasion. It is also smooth, so knots, plugs or wood grain don't show up in the finished wall. According to the APA, 20 uses are not uncommon, and in my experience this is a fair statement. But because it costs 50% more than Plyform, you've got to have a steady diet of upcoming concrete pours to justify its expense.

Formwork lumber should be relatively light, take nails without splitting, be easily worked and readily obtained. Here on the West Coast we use Douglas fir, but in other parts of the country, the typical formwork lumber may be southern pine, eastern spruce and hem fir.

For formwork, #2 lumber is adequate. Its product classification is called "structural light framing." It is cheaper than #1 but relatively close in strength, and it is considerably stronger than #3. Inspect your formwork lumber and reject any that is badly split or has large knots.

Form-tying hardware—In its plastic state, concrete would much rather be a floor than a wall. It wants to escape whatever constrains it and get as close as possible to Mother Earth. The constraint applied by a wall form is supplied by either external bracing or by devices called form ties, which link the opposing forms to one another and prevent them from moving outward. Form ties fit through holes drilled in the forms.

The most common form tie is called a snap tie. It consists of a high-tensile steel wire with a bolt head or button protrusion on both ends. Two plastic cones on the wire, held a fixed distance apart, control the thickness of the wall by bearing against the forms. Snap ties typically come in 2-in. increments. Hence, you buy 6-in., 8-in., 10-in. or 12-in. snap ties depending on how thick you want your concrete wall.

Snap ties can be used as part of a forming method called the single-waler system that includes triangular steel brackets designed to cradle 2x4 walers while holding the end of the snap tie (drawing left). Another U-shaped bracket can be used to fit over pairs of 2xs. They are used to secure double walers and strongbacks.

The bracket designs vary a bit depending on the maker, but they all use some type of a wedge to apply tension to the end of the snap tie. Jahn brackets (Dayton Superior Corp.), for example, are equipped with an eccentric cam for this purpose. The cam has a takeup of ⅝ in. to secure the waler in place and to take into account the variations that occur in lumber sizes.

In the stud-and-waler system, wedges are used to secure the ends of the snap ties. The wedges used with snap ties have a key-hole slot that fits over the end of the tie. As the wedge is driven with a hammer, the head of the tie pushes the wedge against the waler, which in turn pushes the face of the form against the plastic cones. This correctly spaces the two faces of the form and aligns the faces with the walers.

To strip the form, the wedge or cam is driven the opposite direction. Once the forms are pulled, you twist the ends of the ties. Each tie has a nick in the wire near the plastic cones, which causes it to break inside the plane of the wall. Hence the name snap tie. If the tie has a bolt head, you can twist off the ends with a socket wrench. If not, you can use a pipe placed over the end of the tie. Bend it about 75°, then rotate the pipe until the tie breaks. Incidentally, one of the advantages of the bolt head is that you can break off the ties before you strip the panels. This makes stripping easier because you don't have to pull the panel over the tie.

The cones at the end of the ties are free to move outward and to rotate, which allow them to be removed after the form is stripped. The

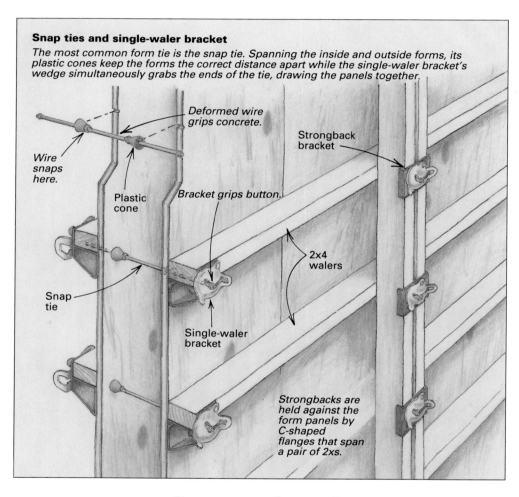

Snap ties and single-waler bracket

The most common form tie is the snap tie. Spanning the inside and outside forms, its plastic cones keep the forms the correct distance apart while the single-waler bracket's wedge simultaneously grabs the ends of the tie, drawing the panels together.

Deformed wire grips concrete.

Wire snaps here.

Strongback bracket

Plastic cone

Bracket grips button.

2x4 walers

Snap tie

Single-waler bracket

Strongbacks are held against the form panels by C-shaped flanges that span a pair of 2xs.

Sources of supply

Atlas SCA, P. O. Box 30, San Diego, Calif. 92112; (619) 277-2100.
Wide selection of hardware geared to commercial or large residential projects.

The Burke Co., P. O. Box 467, Ellicott Station, Buffalo, N. Y. 14205-0467; (800) 423-9140.
Single-waler system. Forming hardware.

Dayton Superior Corp., 721 Richard St., Miamisburg, Ohio 45342; (513) 866-0711.
Jahn single-waler system. Full line of form ties and coil hardware.

Gates and Sons, Inc., 90 South Fox St., P. O. Box 9509, Denver, Colo. 80209-0509; (303) 744-6185.
Single-waler system.

Richmond Screw Anchor, 7214 Burns St., Fort Worth, Texas 76118; (817) 284-4981.
Single-waler hardware, specialty ties geared to heavy construction but of potential interest to the home builder.

RJD Industries, Inc., 26945 Cabot Road, #107, Laguna Hills, Calif. 92653; (714) 582-0191.
Fiberglass form ties.

Williams Form Engineering Corp., P. O. Box 7389, Grand Rapids, Mich. 49510; (616) 452-3107.
Single-waler hardware, specialty ties geared to heavy construction but of potential interest to the home builder.

Alternative form ties

Most residential walls can be cast using snap ties to hold the forms together. But if you spend any time pouring concrete, you're going to run into a situation where snap ties just won't do. The alternatives listed here can solve a variety of forming problems. In addition, concrete-accessory companies make specialized hardware that can be useful to anyone building with concrete (see "Sources of supply" on the facing page).

Pencil-rod ties—You can use steel rods with a special clamp to tie two sides of a form together (top left drawing, below). The rod may be cut to any length and thus used to tie across large distances. I've used pencil rods with good results to tie across the corners of thick walls and to tie forms that vary in thickness, such as those used to cast battered walls. But there are no spacers on the rod to hold the faces of the forms at the proper distance apart. One solution to this problem is to make sleeves out of PVC pipe that fit over the rods inside the forms.

The load capacity of a pencil-rod tie depends on the diameter and the strength of the rod. Typical values are 1,000 lb. for ¼-in. rods to 7,000 lb. for ⅝-in. rods. Check with the supplier for actual values.

The fiberglass tie is a pencil-rod variant that won't rust, so it can be used in walls that will be near corrosive moisture, such as salt water. A fiberglass tie is also nearly invisible when you cut it off flush with the wall, so you don't have to patch holes made by snap-tie cones. RJD Industries makes them.

Taper ties—A taper tie consists of a rod, threaded on each end and tapered along the inside faces of the forms (top right drawing, below). Washers and nuts at each end of the ties keep the form from moving outward. Taper ties are used where it is either necessary to remove the tie from the wall or where the tie spacing produces a high loading on the tie. The tapered portion between the form faces allows the tie to be driven out from the small-diameter end and reused. No spacers are provided with this type of tie, and some other means must be used to hold

the form faces the correct distance apart. Allowable load values range from a few thousand pounds to over 40,000 lb.

She bolts—She bolts are similar to taper ties except that they are externally threaded on only one end. The other end has internal threads (bottom left drawing, below). The bolt tapers toward the internally threaded end. Usually one she bolt is used on each side of the form and connected to a threaded rod inside the wall. The inner rod remains in the concrete.

To remove a she bolt, remove the nuts and unscrew the she bolt from the inner rod. The tapered portion of the bolt breaks free of the concrete almost immediately so that by the time the bolt is unscrewed from the inner rod, there is no longer any resistance. The threaded ties that connect she bolts can be as long or short as required. When she bolts are coupled with short threaded ties, the steel that remains in the wall ends up several inches from the face of the wall, which can be helpful when moisture and

corrosion are a consideration. Allowable load values for she bolts vary, depending on size and range, from a few thousand pounds to over 30,000 lb.

Coil-tie hardware—The principal parts of a coil-thread system are a helically wound coil of wire and a bolt with special threads that grip the coils (bottom right drawing, below). Coil-wire products are quite versatile: If you have to pour a new wall next to an old one, you can grout wire coils into holes drilled in the old wall and use them to anchor the forms in place.

When linked by wire struts, wire coils become form ties. They are placed between the two form faces, and coil bolts are passed through the walers and screwed into the coils. When stripping the forms, the coil bolts are removed, and the tie remains. Some manufacturers make coil ties with plastic cones similar to snap ties so that the coil tie acts as a form spreader too. Different manufacturers use different thread patterns for their coil-wire hardware. They aren't interchangeable. —*W. D.*

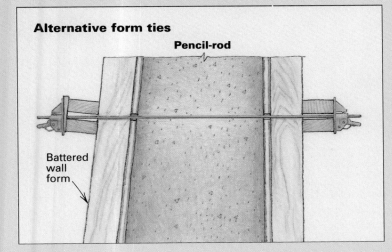

Alternative form ties

Pencil-rod

Battered wall form

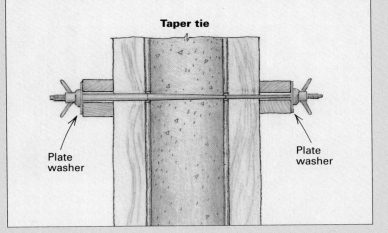

Taper tie

Plate washer

Plate washer

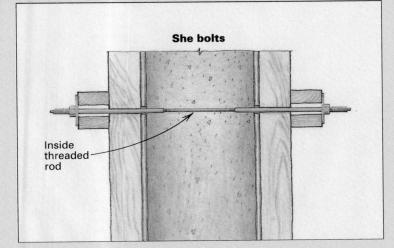

She bolts

Inside threaded rod

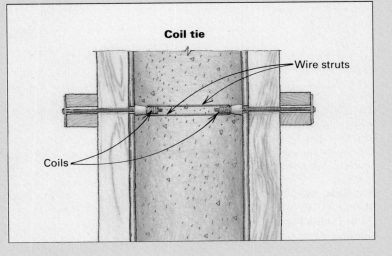

Coil tie

Wire struts

Coils

dimple left by the cones is then patched with grout. Snap ties commonly have an allowable load of 3,000 lb., 4,000 lb. or 5,000 lb.

Snap-ties are fast and easy to install, relatively cheap and self-spacing. On the downside, snap ties are labor intensive when it comes time to patch the holes that they make in the wall. And in wet conditions, the metal left by the remaining portion of the tie can be subject to corrosion—even when it's buried under a grout patch.

For most walls, snap ties are the best choice. But there will be cases where snap ties won't work because they aren't the right length or strong enough. The sidebar on p. 123 describes several alternative ties.

Walkways—We insist on putting walkways on any forms over 4 ft. It's just a lot safer and easier on the crew if they don't have to worry about where they're going to put their feet when they are concentrating on filling the forms.

For walkways, I recommend the prefabricated steel scaffold jacks that are designed to be secured to the walers. The jacks are available from the same companies that make the form hardware, and they can be reused many times.

Resist the temptation to stack heavy materials on walkways supported by scaffold jacks. They are designed to take the weight of a couple of workers—not piles of rebar, blocks of concrete and sacks of cement.

Two form systems—Site-built forms must be easy to assemble and strip; they must not unduly deflect; they must be easy to plumb and line; and they must be safe. The first system that fits these criteria is called the stud-and-waler system (bottom drawing, p. 121). It is made up of panels that are held plumb by studs. Horizontal walers hold the form straight. We typically prefabricate 8-ft. by 8-ft. panels. We use two 16d sinkers top and bottom to secure the plates to the studs, and 8d commons on 2-ft. centers to affix the plywood to the studs and the plates. Running the plywood perpendicular to the studs always makes for a stronger form.

When you put the studs close together, the stud-and-waler system can take a lot of weight. The drawing shows an 8-ft. form that can take a pour rate of 4½ ft. per hour at 70° F. By shortening the distance between studs to 8 in. and shortening the tie spacing to 16 in., the form can withstand full liquid head (see sidebar above). That means the form will contain 8 ft. of concrete that hasn't yet begun to set up.

Our crew likes to use the stud-and-waler system for long, straight runs of wall. The panels can be made up in advance, and the horizontal walers keep the number of ties necessary to a minimum. Ties on 2-ft. centers in both directions are typical for this system. That means less work patching the snap-tie holes. I also think we get a little more mileage out of our materials with the stud-and-waler system because the plates and studs around the perimeter of the panels tend to protect the edges of the plywood, which is where the panels start to deteriorate.

I don't think hand-set form panels should be any bigger than 8 ft. by 8 ft. Anything larger is too

Calculating concrete pressure

Before you can design the forms for a concrete wall, you need to know the maximum pressure that the concrete will exert against the forms. The table below shows in pounds per square foot (psf) how the temperature of the concrete and the depth of the pour affect that pressure.

Fresh concrete approximates the behavior of a true fluid, which means the pressure increases at a constant rate with depth, and the pressure at any point within the fluid is the same in all directions. If the concrete remained fluid throughout the pour, the equation would be simple: the height of the form multiplied by the unit weight of the concrete (150 lb. per cubic ft.). Thus, for a 4-ft. high wall poured in one lift, the pressure at the bottom of the wall would be 4 ft. x 150 lb. or 600 psf. For an 8-ft. high wall poured in one lift, the pressure would be 8 ft. x 150 lb. or 1,200 psf. The pressure exerted on a form filled with fluid concrete is called "full liquid head."

Concrete, however, is not a perfect fluid because, as it passes from the plastic state to a solid state, it starts to support itself. How fast this transition occurs is a function of temperature—the higher the temperature, the sooner the concrete starts to set up.

Other factors that affect the pressure on the forms are the weight of the concrete, how fast it is placed, whether or not the concrete is vibrated in the forms and how far the concrete falls as it enters the forms.

A committee of the American Concrete Institute came up with this pressure formula for walls with a pour rate of 7 ft. per hour or less*:

$$p = 150 + (9,000 \times R/T), \text{ where}$$

p = pressure in lb./sq. ft.
R = rate of placement in ft./hour
T = temperature of concrete.

Say you plan to pour an 8-ft. wall in two lifts of 4 ft. per hour with a temperature of 60°. To find the maximum pressure exerted on the forms under those conditions, first divide 4 by 60. Multiply the answer (.067) by 9,000. Add that number (603) to 150. The answer is 753 lb. per sq. ft.

Now look at the table on p. 120 under "Max. concrete pressure." The fifth entry shows 732 psf, which is within the 5% margin I feel comfortable with. I'll choose the form called for in this column. It says to use 4,000-lb. snap ties on 2-ft. centers, ¾-in. plywood oriented perpendicularly, studs on 12-in. centers and walers spaced 2 ft. apart.

The tables also allow you to work backwards in calculating a pour. For example, if your forms are in place, and the weather turns cold, adjust the pour rate according to the table. Check the concrete's temperature as it comes off the truck, and call your batch plant if you are unsure about the correct pour rate. —*W. D.*

*This formula applies to normal-weight concrete (150 psf) made with type I cement and having a slump of 4 in. Admixtures may increase the pressures.

Placing rate (ft./hour)	CONCRETE TEMPERATURE (F)					
	40°	50°	60°	70°	80°	90°
1	375 psf	330 psf	300 psf	279 psf	263 psf	250 psf
2	600 psf	510 psf	450 psf	407 psf	375 psf	350 psf
3	825 psf	690 psf	600 psf	536 psf	488 psf	450 psf
4	1,050 psf	870 psf	750 psf	664 psf	600 psf	550 psf
5	1,200 psf	1,050 psf	900 psf	793 psf	713 psf	650 psf
6	1,200 psf	1,200 psf	1,050 psf	921 psf	825 psf	750 psf
7	1,200 psf	1,200 psf	1,200 psf	1,050 psf	938 psf	850 psf
8	1,200 psf	1,200 psf	1,200 psf	1,090 psf	973 psf	881 psf

The pressure of concrete on wall forms varies depending on the temperature and how fast the concrete is placed in the forms. Numbers are based on an 8-ft. tall form.

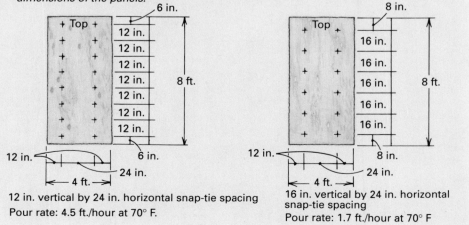

Single-waler snap-tie spacing
Snap ties are typically spaced on centers that are compatible with the 4-ft. by 8-ft. dimensions of the panels.

12 in. vertical by 24 in. horizontal snap-tie spacing
Pour rate: 4.5 ft./hour at 70° F.

16 in. vertical by 24 in. horizontal snap-tie spacing
Pour rate: 1.7 ft./hour at 70° F

heavy. The average person can comfortably lift and carry about 70 lb. A ¾-in. thick sheet of plywood weighs about 72 lb., and if you add a stud frame 12 in. o. c., you've got another 70 lb. or so. Thus one person can carry a full sheet of plywood; two people can carry a 4-ft. by 8-ft. panel consisting of ¾-in. plywood and stud frame; and four people can handle an 8-ft. by 8-ft. panel.

The second system is called the single-waler form (top drawing, p. 121). Unlike the stud-and-waler system, single-waler forms use no nails. The walers hold the form straight, and every 6 ft. a pair of vertical 2x4s, called a strongback, holds the form plumb and keeps the stacked panels in the same plane. With the single-waler system, the plywood panels are stood up vertically, again to keep the face grain perpendicular to the supports, which in this case are horizontal walers. Although the single-waler system doesn't lend itself as well to fast, tall pours, it is easier to set than the stud-and-waler version because there is less lumber involved. And when disassembled, the plywood panels don't take up much room because the lumber isn't nailed to them.

There are two typical tie spacings for single-waler forms (bottom drawing, facing page). Placing ties 12 in. apart vertically and 24 in. apart horizontally will take a pour rate of 4½ ft. per hour at 70° F. If you open up the spacing to 16 in. vertically by 24 in. horizontally, you can pour at 1.7 ft. per hour at 70° F.

Assembly tips—Before erecting the forms, you should spend a couple of hours planning their placement. The goal always is to use as many full panels as possible, and they need to be opposite one another for the snap-tie holes to be aligned. Even if the plan isn't perfect, it will be better than no plan at all. Here are some rules to guide you: lay out the forms on paper first; consider how you will get the rebar into place; cut fillers from full panels and drill the tie holes after they are erected; leave no filler section untied for a distance greater than the spacing of the ties. The typical place to start erecting the panels is at an inside corner (top left photo, above).

Form panels bear against 2x4 plates at the bottom. We affix the plates to the footing by drilling ¼-in. dia. holes every 2 ft. through the plates and into the footing with a hammer drill. Then we drop in a pair of 16d nails—one common and one duplex. They wedge each other in place and work as well as a powder-actuated fastener for forming, plus they can be easily removed.

Once the plates are down, we stack up the plywood panels and gang-drill them for the ties with a ⅝-in. spade bit or auger. We drill four sheets at once, using one of the sheets as a template.

When all the inside panels are raised, braced and affixed to one another with cleats across their tops, insert the form ties and assemble the brackets and the lumber. The inside wall should be substantially complete, with the ties sticking out from the wall. Now hang your rebar and raise the outside forms. Where full panels won't fit in the wall, rip plywood form fillers to fit.

Run the walers long at outside corners and lace them one over the other. Then tie them together with vertical 2x4s (top right photo, above).

Building the forms. Once the plates are affixed to the footing, the formwork panels can be raised. An inside corner is a good place to begin because full panels can be used, and they become self-supporting when tacked together. In the top left photo above, a worker plumbs a panel while the other nails it to the adjacent panel. To brace outside corners, let the walers run wild past the panels (top right photo, above). Then tie them together with duplex nails and vertical 2x4 ties. In the bottom photo above, a worker affixes a turnbuckle form brace to the end of a 2x4. Placing the turnbuckle at the top makes it easy to adjust the form to plumb.

Bracing—Because they are tied to one another internally, these forming systems don't require external bracing to resist the load of the concrete. But they do need braces to keep them plumb and to keep the wind from moving them. Anyone who has tried to pick up a piece of plywood in a strong wind will appreciate the force that the wind can exert on a broad panel.

For forms that are 8 ft. high or lower, setting a braced 2x4 at 45° every 4 ft. should handle the loads imposed by normal wind conditions and the bumps from construction activity. Our superintendent puts strongbacks on single-waler forms at the same frequency to keep them plumb, so they do double duty. We use the turnbuckle-type braces that are affixed to the end of a 2x4 to plumb the forms (bottom photo, above). In dirt, we nail the other end of the brace to a stake. If we're bracing to a concrete floor, we use powder-actuated fasteners or nails in drilled holes to anchor cleats to the floor for the stakes. The forms only need to be braced on one side. □

William Doran is a professional engineer working for Tutor-Saliba-Perini in Los Angeles, Calif.

FOR MORE INFORMATION

WCD-3 Design of Wood Formwork for Concrete Structures, National Forest Products Association, Publications Department, 1250 Connecticut Ave., N. W., Suite 200, Washington, D. C. 20036. $4 plus shipping and handling.

Concrete Craftsman Series-2: Cast-in-Place Concrete Walls ($11.95), *SP-4 Formwork for Concrete* by M. K. Hurd ($100.95), American Concrete Institute, P. O. Box 19150, Detroit, Mich. 48219.

Formwork to Concrete by C. K. Austin, University Microfilms International, Books on Demand, 300 North Zeeb Road, Ann Arbor, Mich. 48106. $81.40 plus shipping and handling.

Cast-in-Place Walls, Portland Cement Association, Order Processing, 5420 Old Orchard Road, Skokie, Ill. 60077-1083. $12.95 plus shipping and handling.

Concrete Masonry Handbook for Architects, Engineers, Builders, Portland Cement Association (address above). $29.50 plus shipping and handling.

Surface-Bonded Block

A strong, fast and inexpensive alternative to poured-concrete or block-in-mortar walls

by Paul Hanke

Pouring concrete walls is a difficult and risky business, and I don't recommend it for the inexperienced. Even professionals sometimes have forms let go, creating various degrees of disaster and pandemonium on the site. Laying up block with mortar has drawbacks, too. It is time-consuming, it takes practice, and the result isn't especially strong.

Surface-bonded block, on the other hand, suits owner-builders to a tee, and can be a less expensive alternative for professionals. It is a method of laying concrete blocks without mortar, then troweling both wall surfaces with a

Paul Hanke is a designer and draftsman at Northern Owner Builder, in Plainfield, Vt.

portland-cement coating laced with chopped fiberglass for strength. Built on standard footings, surface-bonded block walls can be used below and above grade, for foundation walls and for finished living spaces. The method is fast and reliable. It requires no particular skill, and the finished wall is stronger than a block-in-mortar wall.

Surface bonding was originally developed as a low-cost construction technique for self-help housing. A USDA booklet on the subject (Information Bulletin No. 374, now out of print) shows a 12-year-old boy doing a successful job after 15 minutes of practice. Even professional masons are reported to be 70% more productive using this method than laying up block in

the conventional way. The USDA estimates that stacking and bonding 100 blocks would take a person an average of 7.4 hours. Several years ago, two friends of mine, Chapin and Donna Kaynor, built an earth-sheltered house using this technique. It took their crew of four inexperienced people, some of whom worked only part time, less than five days to stack and bond about 1,200 blocks.

Strength and cost—Stacked blocks coated with bonding mix have an average tensile strength (ability to withstand longitudinal stress) of from 300 psi to 500 psi, according to lab tests conducted by the USDA and the University of Georgia. This is about equal to the strength of unreinforced concrete, and is six times stronger than block laid up with ordinary mortar joints. Mortar has very little adhesive power, and virtually no tensile strength. Its main purpose is to level blocks between courses. Because the weakest part of conventional block walls is the bond between block and mortar, these joints tend to crack, making water seepage a problem. A surface-bonded wall, with its seamless outer coating, is much more watertight (though the coating alone should not be relied upon below grade).

Having a block-in-mortar wall built costs about twice the price of the materials, plus footing and reinforcing. In our area, concrete foundation walls currently cost around $95 to $105 per cubic yard poured in place, including formwork and labor. A typical full basement accounts for about 5% of the cost of a house, or over $3,200 for the average $65,000 home. The builder using surface-bonded block can save as much as 35% to 40% of this figure.

Estimating materials—To build surface-bonded walls, you will need standard hollow-core concrete block, surface-bonding mix, galvanized corrugated brick-ties for shims (the Kaynors used about 250 for 1,200 blocks), threaded steel rod and connectors, a few sacks of mortar, and the rebar and concrete for footings. Order 8x8x16 block (about 65¢ each) for walls above grade or foundation walls that will extend less than 5 ft. below grade. Order 8x12x16 block (about $1.05 each) for a foundation wall deeper than 5 ft. Use the USDA table at right to estimate the number of blocks you need. Be sure to add extra block for reinforcing pilasters, the column-like buttresses used to strengthen the walls (discussed below), and half-length block, which you may need for the door and window openings. Order 5% to 10% extra to make up for waste and breakage. Have your block delivered to the center of your work area if at all possible, or deposited in strategic piles around the perimeter. To save time and effort, don't carry those heavy blocks any farther than you have to.

As the table shows, nominal 8x8x16 block is actually 7⅝ in. by 7⅝ in. by 15⅝ in. to allow for ⅜-in. mortar joints; so you can't figure in exact 16-in. modules when laying block dry. Having to calculate with fractional numbers would be a real headache, but estimating tables supplied by the USDA or surface-bonding mix manufacturers greatly simplify the task.

Bonding mix, which comes dry and includes the chopped fiberglass strands, is sold in bags of various sizes. A 50-lb. sack will cover about 50 sq. ft. of wall. Check the exact coverage when you order, and allow about 10% for waste from broken sacks, mixing and troweling. A 50-lb. sack of grey-colored mix currently costs about $14 in Vermont. White is about $17 per sack. You can also mix your own, as explained at right. In addition to the bonding mix, get enough sacks of mortar to lay the first course of block, plus a few extra sacks to use in spots where you need to shim more than ⅛ in.

I recommend using ⅜-in. threaded steel rod to connect the sill or top plate at the top of the wall to the footing below. It is available in 2-ft. and 3-ft. lengths at hardware stores. Threaded rod is expensive, but it makes a secure connection. You'll also need connectors to join the lengths of rod, and nuts and washers to secure the wood sill to the top of the wall. The rod isn't for concrete reinforcement, but to tie footing, foundation and framing together to resist uplift forces. The block cores that contain the rod don't require filling with concrete.

The alternative to running threaded rod all the way up through the wall is to fill the cores at the top of the wall with concrete two or three courses deep every 4 ft., and embed standard ½-in. J-bolts. Use screening to keep the grout from falling all the way to the footing, or stuff fiberglass insulation down the block core. This method works, but it will not provide a continuous connection from footing to sill, and will not resist uplift.

Footings—As a general rule, footings should be twice as wide as the wall above, and as deep as the wall will be thick. A standard 8-in. thick wall calls for a 16-in. wide footing. Pour 24-in. wide footings for either a 12-in. thick wall or a two-story house.

The bottom of the footings should be at least 12 in. below the frost line, and almost anyone can safely pour them. You can pour into shallow forms or directly into trenches of the proper size, provided that their sides and bottoms are of firm, undisturbed soil. Place two No. 4 (½-in. dia.) lengths of rebar near the bottom of a 16-in. wide footing. A 24-in. wide footing will require three lengths of rebar. Check codes for the rebar requirements in your area. Remember to widen the footings for pilasters.

Although you can mix your own concrete for footings, ready-mix concrete delivered to your site is best. Insert the lengths of threaded rod vertically into the concrete at the corners and pilaster locations, on both sides of all the door and window openings, and every 4 ft. to 6 ft. along the wall, as shown in the drawing on the bottom of the next page.

Stacking block—After your footings are poured and have been allowed to cure, you can begin on the walls. Using your batter boards and strings (see "Site Layout," pp. 17-19), drop plumb lines to establish the outside corners. Use the table to determine exact wall lengths, and allow an extra ¼ in. per 10 ft. for irregularities in the blocks. Measure the diagonals to be sure that your corners are square, and adjust if you need to. Snap chalklines from corner to corner as guides, and then lay and level the first course of block in a bed of thick mortar. Check the top of the first course with a 4-ft. level as you go. If a block is too high, tap it down with the butt end of your trowel; if it's too low, remove the block, add more mortar and reset. Don't put mortar in the vertical joints between blocks; just butt them tightly against each other. Some skill is required here. Take your time and do a good job.

The rest of the wall is simply stacked dry in a standard running bond—each block overlapping half the block beneath. Begin by

Mixing your own

- **19½ lb. portland cement (78% by weight),** white or type I grey, which is more common. This is the glue that holds things together. It comes in 94-lb. sacks.
- **3¾ lb. hydrated lime (15%)** for increased workability. It comes in 50-lb. sacks.
- **1 lb. glass-fiber filament (4%),** chopped into ½-in. lengths. Use type E fiber or, better yet alkali-resistant type K fiber, available from plastic and chemical-supply dealers, building-material dealers or boatyards.
- **½ lb. calcium chloride flakes or crystals (2%),** to speed setup time and harden the mix. It's available from agricultural-chemical supply houses. Calcium chloride is also used for salting roads.
- **¼ lb. calcium stearate (1%),** wettable technical grade, makes the mix more waterproof. You can obtain it from chemical distributors.

Since the bonding mix sets rapidly after water and calcium chloride have been added, do not make more than a 25-lb. batch at one time (dry weight).

Begin by mixing the powdered ingredients, except for the calcium chloride. Add the glass fiber, and remix only enough to distribute the fibers well. Overmixing breaks the fibers into individual filaments, which makes application difficult. Be sure to wear a proper respirator. The chemicals are very corrosive, and you don't want to breathe fiberglass, either.

Mix the calcium chloride with 1 gal. of water, and slowly add this solution to the dry ingredients. Mix thoroughly. Add about ½ gal. more water, until the mix is the right consistency—creamy, yet thick enough for troweling. A mix that's too thick is hard to apply and may not bond properly. —*P.H.*

Wall and opening dimensions for surface bonding

Number of blocks	Length of wall or width of openings		Number of courses	Height of wall or openings	
1	1 ft.	3⅝ in.	1	0 ft.	7⅝ in.
2	2	7¼	2	1	3¼
3	3	10⅞	3	1	10⅞
4	5	2½	4	2	6½
5	6	6⅛	5	3	2⅛
6	7	9¾	6	3	9¾
7	9	1⅜	7	4	5⅜
8	10	5	8	5	1
9	11	8⅝	9	5	8⅝
10	13	0¼	10	6	4¼
11	14	3⅞	11	6	11⅞
12	15	7½	12	7	7½
13	16	11⅛	13	8	3⅛
14	18	2¾	14	8	10¾
15	19	6⅜	15	9	6⅜

Blocks sold as 8x16 are actually 7⅝ in. by 15⅝ in. to allow for the size of mortar joints in standard block construction. Remember that cement blocks are not uniform. Add ¼ in. to every 10 ft. of wall length to take this into account, and before beginning to build, make a trial stack to measure the precise height your wall will be.

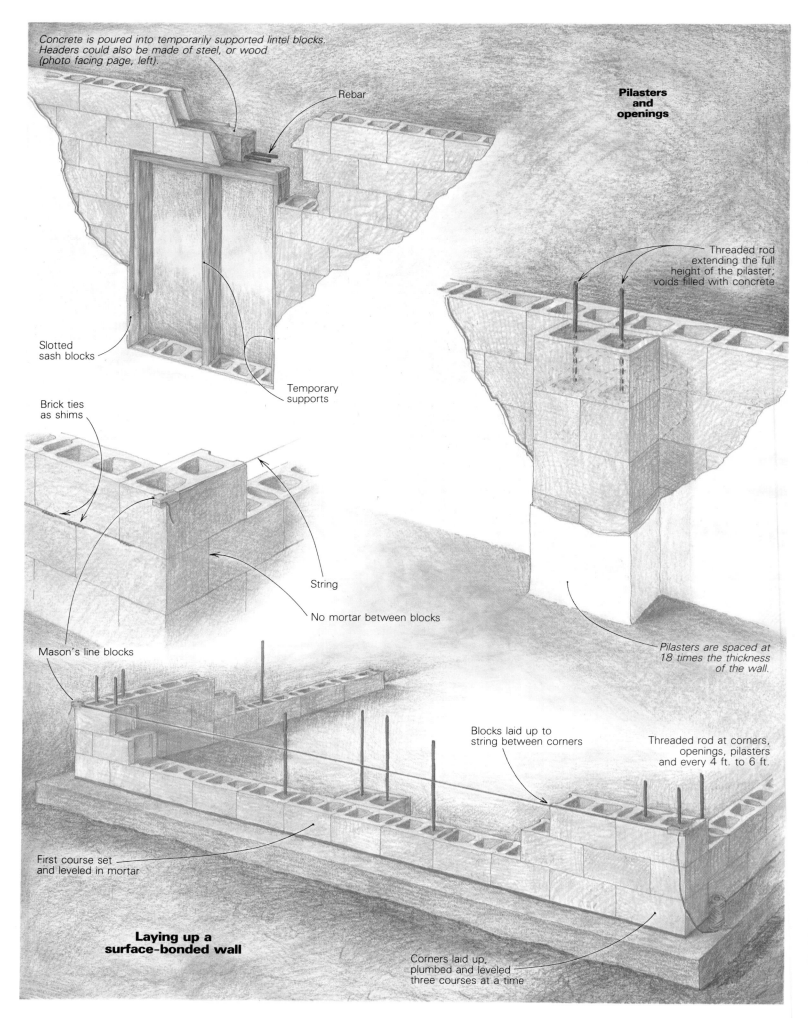

Concrete is poured into temporarily supported lintel blocks. Headers could also be made of steel, or wood (photo facing page, left).

Rebar

Pilasters and openings

Threaded rod extending the full height of the pilaster; voids filled with concrete

Slotted sash blocks

Temporary supports

Brick ties as shims

String

No mortar between blocks

Pilasters are spaced at 18 times the thickness of the wall.

Mason's line blocks

Blocks laid up to string between corners

Threaded rod at corners, openings, pilasters and every 4 ft. to 6 ft.

First course set and leveled in mortar

Laying up a surface-bonded wall

Corners laid up, plumbed and leveled three courses at a time

The first course of block is laid and leveled in mortar, as at right. The corners are built up three courses, then a level line is strung between corners. Dry block fills in up to it. Shims are used where necessary to keep blocks aligned. A mason's corner block, which holds a level line that is also the correct plane for the face of the blocks, is visible on the second course. Once the walls have been built up around openings, headers must be installed. These can be steel, concrete or wood, as above.

building up the corners three courses high. Check them for plumb with the 4-ft. level held vertically, and for level with a water tube. Then stretch a taut string between the top outside edges of each course of the built-up corners. Use mason's line blocks (available where you buy concrete blocks) to secure the line at each end. Fill in the length of the wall up to the string, and repeat the process every three courses, inserting metal shims as necessary to keep the wall level and plumb. If more than ⅛ in. of shimming is required, use mortar instead. Check the wall for plumb at least every three courses. Connect new segments of threaded rod as you go.

Pilasters—These are engaged columns that reinforce the wall against lateral forces and keep it from buckling under heavy loading. For basement walls, pilasters should be on the inside to resist the pressure of the surrounding earth. The Kaynors put theirs outside to get them out of the living area. They are tied in by rotating the blocks of every other course 90° so that they become a part of the wall itself (drawing, facing page). Threaded rod or rebar should extend through the block cores the full height of all pilasters. After the wall is laid up, fill the voids of the pilasters with concrete.

For above-grade construction, pilasters are usually spaced along the wall at a distance equal to 18 times the thickness of the wall (for example, every 18 ft. for a 12-in. wall, or every 12 ft. for an 8-in. wall), or on a shorter wall, at midspan. The pilasters on the house shown here are on 8-ft. centers for earth-bermed walls, which is probably a good precaution for any below-grade construction.

Weight-carrying beams should also be supported by pilasters at each end. Be sure that the beam pockets extend into the wall at least

3 in. to get good bearing surfaces. Once the beam is in place, you can continue dry-stacking blocks in the usual manner.

Openings—For doors and windows, just omit blocks in the proper locations. This is where half-blocks come in handy. With these, you don't need to cut standard block down to size. The blocks at each side should be slotted sash-blocks, which accept a metal or wood spline that attaches to specially made framing.

Headers are required above openings, as in any other type of construction, and they should be properly sized for their span and load. Consult standard tables, codes, or an engineer if necessary. Headers can be made of wood (photo above left), steel angles or U-shaped bond-beam blocks. You support the blocks temporarily over the opening you want to span, then fill their cores with rebar and concrete. Once the concrete has cured, remove the supports, and you have a solid beam.

Coating the wall—Once all the blocks are stacked and the pilasters are filled with concrete, the walls are coated with surface-bonding mix (see sources of supply, below). Commercial mix consists primarily of mortar and strands of fiberglass chopped into ½-in. lengths. Add water according to the instructions, and mix with either a garden cultivator or a mason's hoe (the kind with two large holes in the blade). The mix will cure in about an hour and a half, so don't whip up too much at one time. Hose down the block wall so that the mix won't dry out too quickly, then trowel on a ⅟₁₆-in. to ⅛-in. coat of the paste.

Both sides of the wall get surface-bonded. Use a hawk to hold a comfortable amount of the mix while you work, and press its edge against the wall to limit slop and spilling. Use a

plasterer's trowel, a steel trowel about 12 in. long, and work from the top of the wall down so you can moisten the block as you go if it begins to dry out in hot weather.

The USDA breaks the procedure down into four steps. First, with a series of sweeps of the trowel, spread the mix 2 ft. or 3 ft. upward from the hawk over a section about 5 ft. wide. Then even out the surface by going over the area lightly with your trowel slightly angled. Repeat these two steps over the area just below the block you've just covered, and cover as much area as you can in 15 or 20 minutes. Lastly, clean the trowel in water and retrowel the plastered area with long, firm, arced strokes to achieve a final, smooth surface.

The glass fibers bridge the joints between blocks, and the tensile strength of the wall increases as the concoction cures. Because the fibers are so short, the system won't work if you lay the block in mortar before coating the walls with the bonding mix. The fibers would be spanning almost their entire length, and this would destroy their effectiveness. The interior surfaces of walls can be textured with a light pass with a stiff brush. Mortar pigment can be added during mixing to color the wall. Surface-bonded walls can also be stuccoed or furred out, if you prefer something other than the bonding mix as a finished surface. □

Manufacturers of surface-bonding mix
Fiberbond Surface Bonding Cement: Stone Mountain Mfg. Co., Box 7320, Norfolk, Va. 23509.
Q-Bond: Q-Bond Corp. of America, 3323 Moline St., Aurora, Colo. 80010.
Stack & Bond: Conproco. Box 368, Hooksett, N.H. 03106.
Surewall: W.R. Bonsal Co., Box 241148, Charlotte, N.C. 28224.
Quick Wall: Quikrete, 1790 Century Circle, Atlanta, Ga. 30345.

Forming Concrete Walls

Building forms stick by stick can save on labor and materials

by Arne Waldstein

These days most builders choose to sub out foundation work to a specialist. The foundation contractor, with his crane, modular forms and experienced crew, can accomplish quickly and accurately what a carpentry crew would find to be a labor-intensive and basically unpleasant task. By the time you pay the labor bill, the money saved by building your own forms just might not seem worth the effort. In some cases you might even lose money. On the other hand, when the primary cost of the project is in the materials—in the case of the small builder or owner/builder, for example—building your own forms can be advantageous. If you have more time, energy and friends than cold cash, building forms is the way to go.

Reading Dan Rockhill's article on form-building (see pp. 20-25) took me back to the early '70s when I worked for a company that built banks. We used wood forms for all our foundation walls. Form-building is labor intensive, so it was important for us to minimize the time spent both installing and dismantling the forms. The methods my old crew employed strike me as more efficient than Rockhill's.

Keeping it simple—Rockhill's article explains snap-tie wall construction. The primary difference between his approach and mine is his decision to prefabricate form panels. Nailing together a modular unit, only to tear it apart later, strikes me as inherently wasteful of both time and material. I prefer to install the forms one stick at a time with hardly any nails. This method is quick, easy and yields good results. Heavy lifting is minimized and two people can easily handle the job; in fact, I've written this article assuming that to be the case. (Of course, a larger crew will speed the job considerably, as various steps can be accomplished either si-

multaneously or by teams on the heels of each other.) Just as installation is simplified, the forms also come apart easily with minimal damage to the material. This is significant because the ease with which you can recycle your lumber is an important consideration.

Incidentally, I'd never heard the term "stem-wall" before reading Rockhill's article. I've poured some 4-ft. retaining walls, usually for landscaping purposes, but when it comes to houses, I'm a big fan of full basements. That may be due partly to the fact that I grew up in tornado country, but I also regard the extra space as far too valuable to pass up. Unlike a framed addition, a basement is not something that can be easily added later. If a choice were necessary, I'd skimp on my fixtures or finish materials before I'd eliminate the basement.

I should point out that this is not "my" method in the sense that I developed it. I

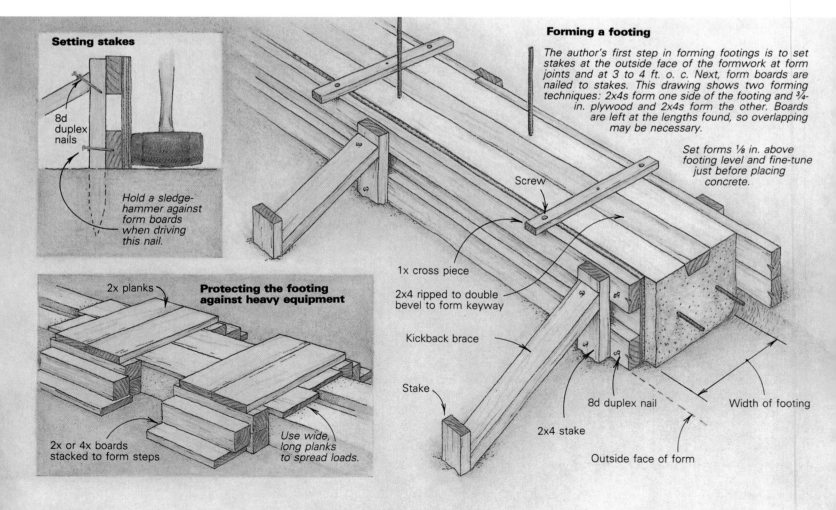

Setting stakes

8d duplex nails

Hold a sledge-hammer against form boards when driving this nail.

2x planks

Protecting the footing against heavy equipment

2x or 4x boards stacked to form steps

Use wide, long planks to spread loads.

Forming a footing

The author's first step in forming footings is to set stakes at the outside face of the formwork at form joints and at 3 to 4 ft. o. c. Next, form boards are nailed to stakes. This drawing shows two forming techniques: 2x4s form one side of the footing and ¾-in. plywood and 2x4s form the other. Boards are left at the lengths found, so overlapping may be necessary.

Set forms ⅛ in. above footing level and fine-tune just before placing concrete.

Screw

1x cross piece

2x4 ripped to double bevel to form keyway

Kickback brace

Stake

8d duplex nail

2x4 stake

Width of footing

Outside face of form

learned the technique from others, and there are undoubtedly many builders out there who are familiar with this approach.

Start with level footings— In building footing forms, use whatever method you are comfortable with. Lapped 2x8s or 2x10s are quick and sturdy. The problems with this approach are heavy lifting (you should avoid cutting this expensive lumber) and the fact that the top edge of the form is usually crowned and sitting higher than you want. This necessitates pouring more concrete or floating concrete to nails or a line snapped on the inside of the form, which slows the pour and is inherently inaccurate. I prefer to use materials that I can set to the correct height and screed off. I like to stack 2x4s edge to edge because they have some lateral strength, yet a slight crown can still be straightened out quite easily (drawing facing page). Also you'll be able to use the 2x4s again. Narrow pieces of ¾-in. plywood and/or 1xs work well, too, although thinner material will obviously require more bracing. Feel free to use anything that is handy, as long as one edge is fairly straight.

I should mention that level footings are even more important if you want to screed off the top of your wall form. Stick-built forms cannot be shimmed as easily as form panels. I prefer to avoid shimming by pouring level footings in the first place. Level footings allow you to simply set the plywood, secure it and screed off the top of the form. Because plywood is a standardized off-the-shelf material, accuracy at footing level will be transferred automatically to the top of the wall. Another advantage of level footings is that the standard 4x8 sheet of plywood can become the unit of reference and measurement in building an 8-ft., or higher, wall. If floating concrete to a line is something you are comfortable with, you can relax on footing accuracy and snap lines at the top of the wall forms.

Achieving level footings is not difficult, but it does require the availability of an expensive precision instrument—the builder's level or transit (see *FHB #37*, pp. 39-45, on how to lay out a foundation). First, lay out your footings by pulling a stringline at the centerline of the wall. Next, lay out stakes and form boards along both sides of the string.

Now set the transit in the middle of your layout and start setting forms. From the stringline, measure a distance equal to half the footing width plus the thickness of the form board and set stakes at this distance, placing them at every joint where the boards meet. Fill in with stakes between joints so that you have a stake every 3 or 4 feet. Pound in the stakes until, as indicated by the transit, the tops are about ⅛ in. higher than the level of the finished footing. Nail and brace the boards as you go. Set the forms slightly higher than the tops of the stakes. Hold the sledgehammer snugly against the inside of the form board as you drive a duplex nail through the stake. That will make nailing easier and more accurate.

I recommend leaving the standard transit rod in the truck. Simply use a clean, straight stick with a single pencil line on it and you'll have fewer lines to decipher. To help you hold the stick plumb, attach a torpedo level to it with a rubber band cut from an old bicycle inner tube. A plumb stick is especially important with respect to leaning toward or away from the transit. Side to side plumb can be recognized by the transit operator as he compares the pencil line to the cross hair.

Once one side of the form is installed, you should make the circuit one more time and set the forms to their precise level. Set the measuring stick on the form itself and pound on the stake. For that last ⅛ in., a standard 16-oz. hammer should be all you need. Now set the other side of the form, using a spirit level. A 2-ft. level with a top-reading center vial is handy for this job. (Check the accuracy of the level before you begin. It is surprising how many people buy levels with adjustable vials, yet never get around to adjusting them. I've seen plenty of arrows and scribbled notes on the sides of levels indicating the "good" side.)

Brace the forms at the stakes with anything that will hold the sides. Perfectly straight sides aren't crucial, so there's no need to get carried away with your bracing. A slight bulge here or an ooze there is no cause for panic. Remember, 30 years from now no one is going to be standing around admiring your footings. Pour the

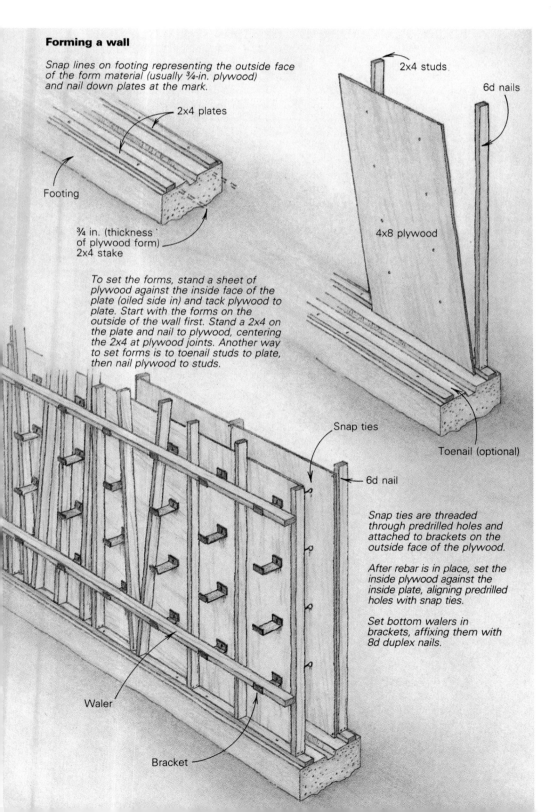

Forming a wall

Snap lines on footing representing the outside face of the form material (usually ¾-in. plywood) and nail down plates at the mark.

2x4 plates

Footing

¾ in. (thickness of plywood form)
2x4 stake

To set the forms, stand a sheet of plywood against the inside face of the plate (oiled side in) and tack plywood to plate. Start with the forms on the outside of the wall first. Stand a 2x4 on the plate and nail to plywood, centering the 2x4 at plywood joints. Another way to set forms is to toenail studs to plate, then nail plywood to studs.

2x4 studs

6d nails

4x8 plywood

Snap ties

Toenail (optional)

6d nail

Snap ties are threaded through predrilled holes and attached to brackets on the outside face of the plywood.

After rebar is in place, set the inside plywood against the inside plate, aligning predrilled holes with snap ties.

Set bottom walers in brackets, affixing them with 8d duplex nails.

Waler

Bracket

Drawings: Christopher Clapp

concrete fairly dry, as you should anyway for maximum strength, and you won't waste much. As long as you have the proper mass and reinforcement, you're all set, and so's the concrete.

Forming a keyway—Rockhill's suggestion that keyways can be eliminated if vertical rebar is embedded in the footing on 4-ft. centers certainly sounds like a timesaver. I prefer keyways, however. Put a slight angle on a table saw and rip 2xs into strips with double bevels to allow easy removal from the footing (drawing, p. 130). After laying the reinforcing steel, nail the keyway strips to 1x crosspieces slightly longer than the width of the formed footing. Then attach the crosspieces to the form using screws and a screwgun—a good way to attach the material without pounding the forms out of level. After you've poured and screeded the footing, gently install vertical rebars next to the keyway form on alternating sides.

One technique I do not recommend for making keyways is jabbing or dragging a board through the wet concrete after it has been screeded. I've seen this done many times, but you will have wasted your earlier precision because the concrete will bulge all over the place. If you are saving your precision for later, then by all means, jab away.

Forming the walls—After you've stripped the footings and saved the wood, you're ready to build walls. Before you start on the walls, however, you might want to check for the proper grade of the basement. If you have a lot of material either to add or to remove, now's a good time to do it, while you can still get a machine in. It is a wonderful feeling to lean against your shovel and watch a machine move around gravel or sand that you would otherwise be shoveling by hand. Be sure to protect the footings where the machine must drive over them. Lay planks on top of the footing and build up both sides with timbers or more planks (drawing, p. 130). Laying plywood on top of that will also help distribute the load. You may have to temporarily bend over a couple of the vertical rebars.

The first step in building walls is to spray the plywood with form oil. Don't wait until the forms and steel are installed, as it is slower and you risk oiling the rebar and the top of the footing. As you oil each sheet, stack it neatly face to face, corners and edges perfectly aligned. Then, using a long bit or an extension, drill all the snap-tie holes at once. No jigs are necessary; simply mark the top sheet and clamp it to the sheets below. Of course, you can drill first, then oil if that's more convenient. If you haul the plywood on your truck, you can drill the stack right on the truck and oil them as you unload them.

Next, determine the width of your concrete wall and snap lines ¾ in. wider (that's the thickness of your form material, usually ¾-in. plywood) to indicate the location of the plates. Nail down some straight 2x4 plates with concrete nails (drawing previous page). Now you're ready to start setting forms. My example assumes a wall height of 8 ft., but that can easily

be varied with no basic change in technique. Eight feet is handy because that's how tall plywood is. But if, for example, you want a 10-ft. high wall, you can rip plywood in half lengthwise and run those 2-ft. pieces horizontally above the lower plywood forms.

To set the forms, stand a sheet of plywood up against the inside edge of the plate, oiled side in, and tack the bottom corners of the plywood to the plate with 6d nails (drawing previous page). Now stand a 2x4 on top of the plate with half of one edge abutting the plywood, and nail the plywood to the 2x4 with two more 6ds, one at the top and one at the bottom. If the stud is really crooked, you can straighten it with another nail about halfway up. You can lightly toenail the stud to the plate if you wish, but it isn't necessary. In fact, if you set the stud first and toenail it to the plate (6ds work fine), you don't really need to nail the plywood to the plate. Just stand it up and tack it to the stud. In either case, if you're not working in a windstorm, the plywood will stand by itself. If you opted for "quick and dirty" footings, you must plumb the first sheet of plywood to keep your seams tight.

Now simply repeat this procedure. Every four feet you will have a plywood seam backed by a 2x4. Plumb and brace the whole thing every so often. You can forget about all that cutting, measuring and nailing of studs to plywood at 16 in. o. c. You'll be doing enough of that ritual once you're above ground, but it's not necessary for formwork.

Once all the outside forms are up, you are ready to hang the snap ties and steel. Refer to Rockhill's article for good rules of thumb for reinforcing steel and an illustration of a snap-tie rod. My experience is with U-shaped brackets, rather than with the wedges shown in Rockhill's drawing.

I prefer to install the forms for the outside of the foundation wall first because it is generally easier to install the rebar from the interior of the basement excavation where there is plenty of room, rather than stumble around at the edge of the embankment. One person pokes a snap tie through each hole in the plywood and another person hangs a bracket on the snap-tie stem. With all the snap ties in place, lay and tie the rebar.

With the steel in place, set the inside plywood on the inside of the plate, aligning the predrilled holes in the plywood with the protruding snap-tie stems. This time around, because you can't really stand inside the wall cavity and nail through the face of the plywood, you'll have to toenail one of the plywood bottom corners to the plate from the back side. You will also need a stepladder in order to nail the top corners to the studs. As a practical matter you don't even need to nail that lower inside corner. As long as the plywood is sitting securely on the footing and the seam is tight, the other three nails will do the job. (Those of you with arms 8 ft. long can go ahead and nail that inside corner.) Hang and secure the brackets on the face of the form as you go, and that will hold everything together.

With all the plywood up, your next step is to set the bottom walers in the brackets and begin setting studs (drawing previous page). The brackets have holes in them that allow you to secure the walers with 8d duplex nails. Set the studs against the plywood and add as many additional studs as you need. With the bottom waler in place, the studs will stand there all by themselves. There is no need to nail anything here.

One letter in *Fine Homebuilding* complained of bulging forms when Rockhill's method was used to form an 8-ft. wall, and Rockhill responded that his stemwall forms must be redesigned for a taller wall. Well, the commercial walls I've built were between 9 ft. and 10 ft. high (to allow for sizable ductwork and a suspended ceiling) and the only "designing" I ever did was to add some extra studs. You can space them 6 in. o. c. if it makes you feel comfortable (and you have enough lumber). The point I'm trying to make is that because beefing up the form doesn't involve any additional work, such as measuring, cutting and nailing, thus creating even more work at the dismantling end, there is no reason not to err on the safe side. Pack in lots of 2x4s. When all the studs are in place, install the rest of the walers.

An important point to note here is that there is no need for a top plate. It is wasteful to needlessly cut your lumber, so put a waler near the top and let the various length 2x4 studs you have on site protrude above the form. Even on a short wall, you can use 8 ft. and longer studs. Use your tallest sticks on the inside in order to allow room for the concrete chute on the outside. When it's time to pour, drive the truck parallel to the wall. Cut a block that can slide between the studs and ride on top of the plywood to screed off the top of the wall. After screeding, install the anchor bolts. One last thing: the absence of a top plate may cause the plywood to undulate a bit, but relax—the weight of the wet concrete will straighten it out in a big hurry.

Stripping the forms—The beauty of this system becomes apparent when it's time to strip the forms. Pull the duplex nails and remove the walers. When the last walers come off, your studs will simply drop to the ground (watch your head!). Your lumber is uncut and ready for action. The seam studs easily pry away because they are only held in place by a couple of 6ds. Now remove the bottom plates (because they are nailed to the footing, this might require a good lever and some elbow grease). That leaves only the plywood, which normally adheres slightly to the concrete. Actually, that is rather convenient, because as you pull the plywood off, one sheet at a time, you can clean it and stack it without having a mess of plywood all over the ground. (Of course, if you didn't oil the plywood adequately, concrete may adhere more than slightly.) Stack your lumber, and when the concrete cures you'll be ready to start framing the walls. □

Arne Waldstein is a builder in Great Barrington, Massachusetts.

Building with Pumice-crete

This lightweight, insulating concrete holds up the roof while holding down the costs

by Vishu Magee

When my wife and I started to plan our own house, we bumped up against the same dilemma faced by many people in the Southwest: We wanted the massive masonry walls and soft curves of adobe, but we couldn't afford adobe. So I decided to investigate building the walls out of lightweight concrete, using pumice as the aggregate instead of sand and gravel. The house we built in Taos, New Mexico, using the lightweight concrete fits right into the state's earth-building tradition (bottom right photo below).

Pumice is a glassy, lightweight igneous rock that reveals its unusual structure under the microscope: It's mostly air. Formed as expanding gases bubble through cooling lava, pumice is light enough to float on water yet sturdy enough to serve as the aggregate in load-bearing walls. Its honeycomb matrix of air pockets also makes it a good insulator. Pumice is found wherever there has been volcanic activity, such as here in New Mexico and in the Pacific Northwest.

In New Mexico, pumice has been used extensively as a building material since the 1970s. Roofers first started spreading layers of dry pumice a foot thick atop flat roofs, sloping it toward the scuppers to create the necessary runoff. With an R-value of 2.25 per in., the pumice serves as both a substrate for the roofing membrane and a layer of insulation.

Around the late 1970s, owner/builders living in and about Taos started experimenting with pumice as a lightweight aggregate in concrete. They found that pumice-crete can be inexpensively formed into walls that, once plastered, are dead ringers for adobe. In 1980, Scott MacHardy of Salida del Sol Construction began researching and developing pumice-crete as a recognized building system. The credit for pioneering this new technology goes to him.

Initially, New Mexico approved pumice-crete as an infill material only—okay for walls in a post-and-beam structure of concrete or timber but not for walls carrying roof loads. Eventually the material was certified as a primary material in walls at least 12 in. thick, and now pumice-crete is also accepted for two-story buildings. Pumice-crete walls are rated at 350 psi to 450 psi, which easily complies with the 300-psi rating required for adobe by New Mexico's building code.

Advantages—Pumice has several qualities that make it a compelling alternative to other earth-masonry construction methods. It combines

Curves are easy. Asphalt-impregnated fiberboard bends to the task of forming the author's dining-room wall (photo above). When the pour reaches the height of the outside form, the rough window bucks will be carefully positioned and held fast against the inside form while the pumice-crete sets up. The rough texture of the pumice-crete is revealed once the forms are stripped away (photo bottom left). Its pebbly surface makes an excellent ground for the layers of pigmented stucco that finish the exterior of the house (photo bottom right).

high insulation values (R-1.25 to R-1.5 per in. when used as aggregate) along with thermal mass. This integral insulation eliminates having to apply a foam layer to the exterior, as is typical in adobe construction. And the rough surface of a pumice-crete wall (bottom left photo, p. 133) is a perfect ground for stucco and plaster, eliminating the need for the stucco netting that is required over adobe block.

Pumice-crete weighs about 45 lb. per cubic ft. compared with regular concrete at 150 lb. per cubic ft. That's partly because there's a lot less water in a batch of pumice-crete than in a similar volume of conventional concrete. As a result pumice-crete will partially support itself in the forms, and that translates into minimal formwork (photos right). The price for poured pumice structures here is about $5 per sq. ft. less than adobe buildings.

Formwork—Straight walls are formed between panels made of 4x8 sheets of ½-in. plywood braced with 2x4s on edge. The height and length of the forms never falls on such a convenient module, so it's handy to have a selection of 2 ft. x 8 ft. and 4-ft. sq. panels on hand for extending the forms as necessary.

The forms bear on footings or a slab and are held plumb with a diagonal brace at each joint. Their bases are secured with 2x4 braces and stakes, and opposing forms are linked at the top with wood cross ties on 4-ft. centers. Abutting forms are joined with 16d duplex nails on 2-ft. centers.

The interior forms go up first and act as a backdrop for positioning windows and doors. Each window and door is installed inside a heavy wood frame called a roughbuck, which typically consists of 2x6 or 2x8 side jambs, a sill and a solid or built-up lintel (drawing facing page). A 2x4 bristling with 16d duplex nails on 4-in. centers is attached to the back of each roughbuck jamb to anchor the frame to the pour. Expanded metal lath stapled to the edges of the roughbucks actually serves as a form for the pumice-crete, creating the inward curve around window and door openings. The metal lath also reinforces the plaster and the stucco that will later wrap around the edge of the roughbuck.

The roughbucks for the doors are placed and temporarily affixed to the interior forms by way of screws and blocks to get the doors aligned correctly. Meanwhile, window locations are noted with chalklines on the inside face of the forms. The window bucks aren't installed until the first lift of pumice-crete has been poured.

Before the exterior forms go up, electrical and plumbing runs must be considered, along with the placement of interior partition walls. We tack 2x2s inside the forms to act as blockouts for wiring raceways. Plumbing runs that are going

Straight walls. **Rudimentary plywood panels stacked on one another are all it takes to contain pumice-crete. In the top photo, the second lift has been poured, and now the bond-beam steel is in place. Strips of extruded polystyrene and Celotex tacked to the inside of the forms act as screeds. In the bottom photo, the first pour stops at the roughbuck's windowsill.**

to be buried inside the walls are installed and tested before the exterior forms go up. Finally, we tack 2x4 nailers (again with duplex nail anchors) to the interior forms wherever cabinets require an anchorage or a partition wall is to engage a pumice wall.

Walls without windows get full-height, exterior-side form panels. Walls with windows, however, have 4-ft. tall forms, allowing the window bucks to be installed once the first pour reaches the correct height (usually 3 ft. above grade).

Pumice-crete walls range from 12 in. to 20 in. thick but are commonly 15 in. thick. The distance between forms is regulated with 1x2 spacers at the bottom and with baling wire ties run through holes in the forms on 24-in. centers.

Because formwork can be minimal, pumice-crete offers a spectacular advantage if you want to build curved walls. To form our curved dining room, we bent ½-in. Celotex (asphalt-impregnated fiberboard) to the desired radius and braced it with 2x4s on 16-in. centers (top photo, previous page). The curve at the bottom of the forms was controlled by wooden blocks affixed with concrete nails to the top of the foundation. The blocks remain in the wall.

The first lift—Pumice-crete has to be mixed with uncommon care—don't even think about souping up the mixture to get it to flow. Too much water can wash out the cement, resulting in poor adhesion in the upper wall and too much density in the lower section. Here are the ratios we use: 1 yd. of pumice screened to ⁵⁄₁₆-in. aggregate; two sacks of type I or type II portland cement; and 19 gal. of water. The correct texture is dry for concrete, resembling raisins covered with gray chocolate. Mix everything on site if you are in doubt about the reliability of your transit-mix supplier.

You can't pump pumice-crete because it isn't a slurry. A backhoe most efficiently conveys the mix to the top of the forms, where a worker directs its flow with a shovel. On sites that are too steep or too heavily forested to maneuver a backhoe, a crane with a bucket is the best, albeit expensive, way to go.

The first lift of pumice-crete stops 4 in. shy of the window-buck bottoms. It doesn't take much vibration to move the pumice-crete around—just some poking with the shovel and banging on the form sides.

Window bucks should be installed on top of a 4-in. thick, high-density, reinforced pumice-crete subsill (drawing facing page). A yard of high-density pumice-crete has 500 lb. of sand and an extra sack of cement. The subsill includes a couple of #3 reinforcing bars that extend at least 4 in. beyond the window jambs. The sills are cast on top of the fresh pumice-crete, and the bucks are then set into the wet, high-density mix, plumbed in both directions, and blocked and braced with screws run through the interior forms. The subsills help to spread out the loads concentrated at the window corners, thereby minimizing cracking at these vulnerable spots. When the first lift of pumice-crete is complete, we cover it with plastic to slow evaporation. Then, where necessary, forms are added to get the outside walls to their full height.

The second lift—Once all the forms are erected, chalklines are snapped on their inside faces to locate a 6-in. deep bond beam. The bond beam is a standard high-density, reinforced concrete layer atop the walls that encircles the building, tying the walls together and distributing roof loads. Because it is cast of standard concrete, the bond beam is insulated by a band of 2-in. extruded polystyrene tacked to the exterior form. A band of Celotex on the interior form acts as a depth gauge and screed for the bond beam. You know the second lift is high enough when it reaches the bottom of these bands.

Ideally, the second lift should happen the day following the first pour. The bond-beam rebar should be installed immediately, and if time permits, we pour the bond-beam concrete as soon as possible to avoid a cold joint with the pumice-crete. The roof structure bears on a pressure-treated wooden plate, which is secured to the bond beam with anchor bolts on 6-ft. centers.

Parapets protect the ends of the roof framing members and conceal the roof membrane, the plumbing stacks, the fans and the skylights behind a finished edge. You can make the parapets out of pumice-crete, but I don't. They're usually beyond the reach of the backhoe, they don't need to be insulated, and they're easier to frame with wood (drawing right). As with other wood-to-masonry joints, the intersection should be bridged with expanded metal lath.

Finishing—Freshly stripped and still green, pumice-crete walls are easy to carve with a mud hatchet (a tool used for working adobe). And even after the walls are fully cured, the material can be excavated fairly easily if plumbing and wiring channels have to be added. Before plastering the walls, we cover the channels with expanded metal lath attached to the walls with 16d nails.

Even in New Mexico it takes a long time for the moisture to leave a pumice-crete wall. So we delay plastering until at least a month after the pour—three months is even better.

Because a pumice-crete wall is so porous, water passes right through it. This won't hurt the structure, but it's rough on interior finishes. That's why it's imperative to wrap the exterior face of pumice-crete walls with a durable stucco skin. I've had good luck with Sto Flex, an elastomeric stucco finish that contains acrylic fibers (Sto Industries, P. O. Box 44609, 6175 Riverside Dr. SW, Atlanta, Ga. 30336; 800-221-2397). I specify a two-coat gypsum plaster for the inside walls.

Wall sections below grade must be stuccoed and sealed with a foundation coating. Some builders use liquid asphalt as a sealer, but in this vulnerable area I combat moisture with a waterproof membrane, such as Bituthane (W. R. Grace Co., 62 Whittemore Ave., Cambridge, Mass. 02140; 617-876-1400). The joint between the wall and the footing should be brushed clean and sealed with a compatible caulk.

Problems and solutions—Like any building system, pumice-crete has its weak points. Here are some lessons that have been learned in the first 10 years of this infant technology.

A proper bond between pours is essential to avoid plaster and stucco cracks as the building shrinks and ages. Leaving the top of a pour jagged will help the next lift lock in, and covering each lift with plastic will help the mix to cure slowly. Forms should be left on for two to three days to promote a good cure. And you should avoid pouring in temperatures below freezing.

Garden walls or parapets made with pumice-crete should be capped with 4 in. to 6 in. of high-density pumice-crete, followed by a layer of Bituthane or single-ply roofing membrane. This step will minimize damage from freeze-thaw spalling at the top. □

Vishu Magee lives in Taos, N. M., and designs homes in the Taos/Santa Fe area. Photos by the author.

Drawing: Christopher Clapp

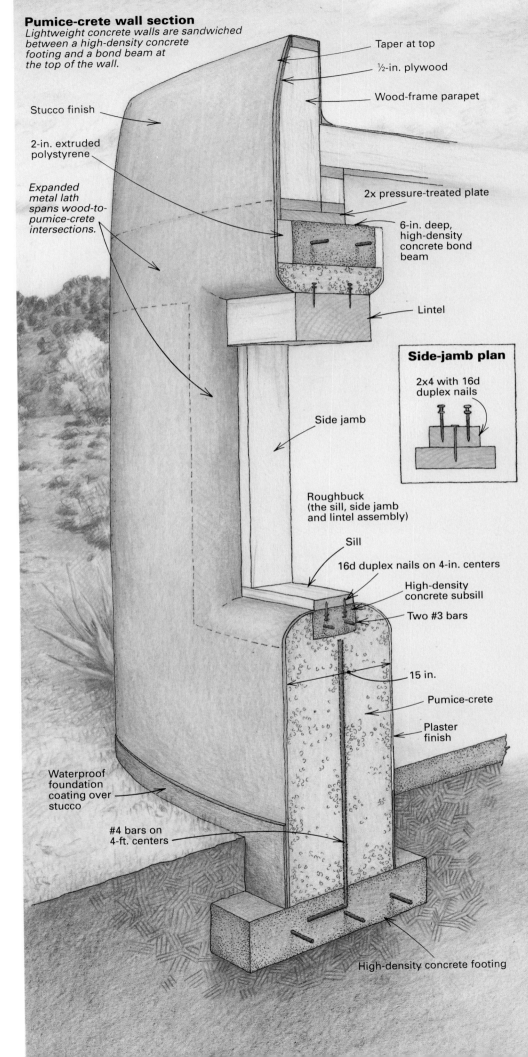

Pumice-crete wall section
Lightweight concrete walls are sandwiched between a high-density concrete footing and a bond beam at the top of the wall.

Taper at top

½-in. plywood

Wood-frame parapet

Stucco finish

2-in. extruded polystyrene

Expanded metal lath spans wood-to-pumice-crete intersections.

2x pressure-treated plate

6-in. deep, high-density concrete bond beam

Lintel

Side-jamb plan

2x4 with 16d duplex nails

Side jamb

Roughbuck (the sill, side jamb and lintel assembly)

Sill

16d duplex nails on 4-in. centers

High-density concrete subsill

Two #3 bars

15 in.

Pumice-crete

Plaster finish

Waterproof foundation coating over stucco

#4 bars on 4-ft. centers

High-density concrete footing

Placing a Concrete Driveway

Careful preparation of the soils and proper positioning of joints and reinforcement are keys to a driveway that lasts a lifetime

by Rocky R. Geans

An attractive, long-lasting alternative to asphalt. Properly installed, a concrete driveway will stand up to lifetime of vehicular traffic with minimal maintenance.

A few years back, we had just finished a driveway at a house in South Bend, Indiana. We were about to call it a day when the homeowners' dog got loose, burst into the garage and used our freshly placed concrete drive as his escape route. When the dog heard us holler, he stopped in the middle of the driveway. Then we tried to get him back inside, which made him scamper back and forth on top of the concrete, making a bad situation worse. We ended up refinishing the whole driveway and giving that dog a scrubbing he'll never forget.

A driveway is only as good as the soils beneath it—We've been putting in concrete driveways for 30 years, and a crucial part of proper driveway design is making sure the materials below the concrete are adequate (drawing facing page). The first 6 in. to 8 in. of material directly below the concrete is the base. The subbase is the soil 8 in. to 12 in. below the base, and the subgrade is usually the native or naturally occurring soil below the subbase. The design thickness of each layer depends on the soil being built on. Acceptable natural soils such as sand and gravel let moisture drain. If the subgrade consists of this type of soil, then it can be compacted to serve as the base and subbase, and more excavating and filling are unnecessary.

However, if the subgrade is clay, peat or fine-grained silty soil that holds moisture and drains

Recycled concrete forms the subbase. Coarsely crushed concrete is compacted to provide a drainage layer and a capillary break to prevent moisture from wicking up.

poorly, removal of up to 20 in. of subgrade soil might be necessary, depending on the support value of the soil. If you have doubts about the soil characteristics in your case, it's worth hiring a soils engineer to do an evaluation.

Establish driveway elevations early for proper drainage—Prior to excavating and backfilling, the exact elevation of the top of the drive should be established. Then, as earthwork is being done, base grades can be brought up with equipment usually to within an inch of their required height, which saves on hand-grading later. For the best drainage, we try to drop the driveway at least ¼ in. per running foot away from the house.

Some situations prevent proper drainage, such as an area of concrete that is locked between a house and a garage. In these cases a catch basin may have to be installed as part of the driveway's drainage system.

The best way to remove water from a catch basin is to use a drain pipe at least 4 in. in dia. that returns to daylight or to a storm sewer that is located safely away from the house. A second method is connecting the catch basin to a dry well. In the most extreme cases, a sump pump is installed to pump collected water to a safe place. The last solution is the most costly and probably should be used only with the recommendation of an engineer.

Soils must be compacted to support the driveway—I check soil compaction initially by walking over the area to get a feel for firmness. Additionally, I shove a ¼-in. to ½-in. dia. smooth steel rod into the soil in several places to check the resistance. If the soils are properly compacted, the rod should encounter firm, even resistance over 2 ft. to 3 ft. However, if the rod meets resistance, say, in the first 6 in. and then can be pushed farther into the soil with ease, it's a sure indication that only the top 6 in. of soil is compacted and that the lower layers of soil are loose. Over time, loose, uncompacted soils will settle as storm water drains through them. Soil settling leaves voids that greatly increase the odds for driveway cracking, sinking or even collapsing in certain areas.

If testing reveals uncompacted soil layers below a top layer that is compacted, then the top soils need to be removed and lower levels compacted properly before the base layers are replaced. In many instances the top 12 in. may be properly compacted while the next 3 ft. to 4 ft. are loose. Ideally, the soils should be excavated down to solid native soil (usually no more than 4 ft.). The soil is then replaced and compacted in 6-in. layers called lifts.

In new construction where excavation has occurred for a foundation, soils should be backfilled and compacted in 6-in. lifts all of the way to final grade. Otherwise, concrete work such as driveways, sidewalks and patios will settle over time and slope or fall inward toward the house. If there are doubts about compaction in any situation, some companies perform on-site compaction tests, a minor investment that can buy peace of mind.

In addition to being properly compacted, the subbase and base need to be made out of materials that will form a capillary break. A capillary break is a layer of soils large and coarse enough to prevent water from being drawn up into them through capillary action the way lamp oil is drawn through a wick. Moisture allowed to wick up and accumulate in soils beneath the driveway slab will freeze, expand and create frost heaves in the slab. Zero-in. to 1½-in. stone or recycled concrete for a subbase compacts well (photo right, facing page) and creates a good capillary break.

The forms determine the final grade of the driveway—After the soils have been layered and compacted satisfactorily and the proper grades established, the driveway forms can be laid out and installed. We usually make our forms out of 2x4s if the driveway is to get a 4-in. slab or 2x6s for a 6-in. slab. The forms need to be staked strongly enough to hold the concrete and to withstand screeding without movement, so we drive our stakes every few feet. Wooden stakes are okay, but they don't hold up well to

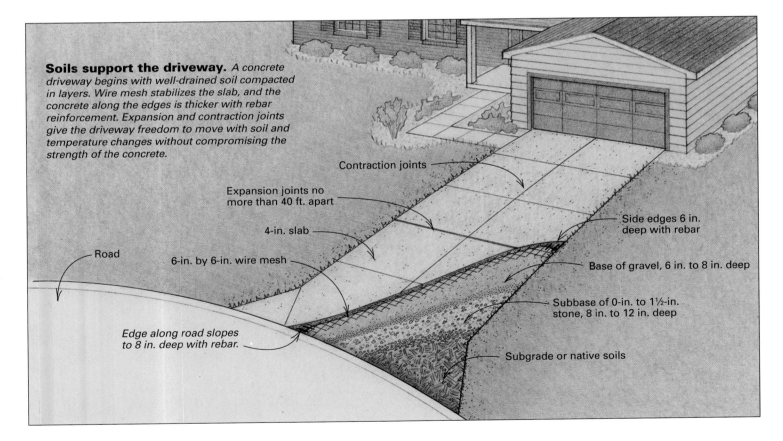

Soils support the driveway. *A concrete driveway begins with well-drained soil compacted in layers. Wire mesh stabilizes the slab, and the concrete along the edges is thicker with rebar reinforcement. Expansion and contraction joints give the driveway freedom to move with soil and temperature changes without compromising the strength of the concrete.*

Contraction joints

Expansion joints no more than 40 ft. apart

4-in. slab

Road

6-in. by 6-in. wire mesh

Edge along road slopes to 8 in. deep with rebar.

Side edges 6 in. deep with rebar

Base of gravel, 6 in. to 8 in. deep

Subbase of 0-in. to 1½-in. stone, 8 in. to 12 in. deep

Subgrade or native soils

Drawing: Vince Babak

The base layer is graded to the proper height. The laser level in the background is used to set the forms to the correct height. The crew then grades the base to the forms, and the base gets compacted one last time.

Expansion joints get a plastic cap for installation. Expansion joints act as buffers between driveway sections. After the concrete has cured, the plastic cap is removed, and sealer fills the joint to keep out moisture and light.

Wire mesh is pulled up on top of a thin layer of concrete. A crew member pulls the wire-mesh reinforcement up to the proper height with a rebar hook. A bead of concrete laid first supports the mesh during placement.

Screeding cuts concrete to the right height. A long, straight 2x4 or aluminum box beam is pulled across the surface of the concrete while it is moved in a side-to-side motion. Screeding brings the level of the concrete down to the top of the forms and consolidates it at the same time.

being driven and generally can be used only once. On most jobs we use commercially available round metal stakes with predrilled holes for nailing the stake to the form.

The driveway featured here was placed on a nice, gently sloping lot. We began by setting the forms to the natural grade along one side. On the opposite side we ran a string at the exact width of the finished slab. We staked the forms for this side along the string at roughly the correct height. Next, we leveled from one side to the other to set the exact height of the opposite forms, driving the stakes deeper or pulling them up to adjust the height. We've found that a laser level is the quickest and easiest tool for setting the height of our forms, although a transit or even a water level can be used.

Our forms usually receive a coating of release agent (a special oil available from concrete-supply houses) to provide better consolidation of the adjacent concrete and to make removal easier. After the forms are set, we grade the base layer of soil to its exact elevation (photo top left) and run the compactor over it one last time. Then we dig the edges of the driveway down a couple of inches. The thicker concrete along the perimeter provides additional support as well as protects against erosion of the soils next to the drive. We also incorporate a double run of rebar along the edges for additional support.

Expansion joints allow the concrete slab to move with changes in weather—Another crucial part of the driveway layout is planning for contraction and expansion joints. Contraction joints are added during placement, so I'll discuss them later. Expansion joints, installed

before concrete placement, allow the driveway to move both horizontally and vertically. Most people think concrete is solid and unmoving. However, concrete not only moves in relation to other solid structures, such as foundations and roadways, but it also expands and contracts with temperature changes and moves as soil conditions beneath the slab change.

Expansion joints provide a full division between different sections of concrete placement. For the driveway featured in this article, expansion joints were placed between the driveway and the sidewalk to the front door, between the driveway and the garage apron, and between the two main-driveway slabs that were placed or poured separately. We didn't need an expansion joint between the driveway and the asphalt roadway, but a joint is required if the roadway is concrete. An expansion joint consists of a thin layer of energy-absorbing material such as asphalt-impregnated fiberboard, plastic foam, wood, cork or rubber. For most driveways, we use ½-in. thick fiberboard installed with a plastic cap strip on top (photo center left, facing page), flush with the finished height of the slab. After the concrete has set up, the cap strip is removed, and we fill the top of the joint with a joint sealant, SL-1 (Sonneborn, Chem-Rex Inc., 899 Valley Park Drive, Shakopee, Minn. 55379; 800-433-9517), which protects the joint from moisture penetration and UV-degradation. The joint sealant also matches the color of the concrete to add a more pleasing look to the joint.

Expansion joints serve several functions in a concrete driveway. First, the expansion joint provides relief between slab sections as the concrete expands and contracts with temperature changes. This movement is horizontal, and the joints that serve this function should be placed no more than 40 ft. apart in any direction. To facilitate this horizontal movement, the base that rests below the slab has to be smooth and well-compacted, with no obstructions such as rocks or holes, which can fill with concrete and restrict movement.

The concrete slab needs to be able to slide back and forth over the base. Any restriction of this movement will contribute to cracking. Even the thickened edge at the roadside is designed with a gradual slope from 4 in. to 8 in. rather than having an abrupt 90° excavation that would restrict slab movement.

Expansion joints also serve as a buffer between the driveway slab and the adjoining rigid structures, in this case the sidewalk and the garage. Movement at these points is vertical; the driveway moves in reaction to changes in the soils beneath the slab. An example is the soil swelling that occurs when moisture in the slab freezes. Without the aid of the expansion joint, the slab would chip or crack as it slid by slight imperfections in the abutting concrete.

Another way we restrict vertical movement is by drilling and installing steel dowels into adjoining rigid cement work, such as a foundation just below the driveway slab. A ⅜-in. by 12-in. dowel works well in most situations.

The foundation, which is bearing on a footing below frost level, should not move. However, the driveway, which rests on soils only 4 in. to 6 in. below grade, will almost certainly experience some degree of movement. We want to give the drive the freedom to move up but not to drop any lower than the dowel.

We also use dowels to maintain alignment between the driveway and existing flat work, such as sidewalks. The dowel should be smooth and installed parallel to the concrete, not at an angle. Rebar can be used for this application, but it won't work as well because of its rough texture. We drill a hole into the existing flat work slightly larger than the diameter of the dowel, which allows for expansion and contraction. These dowels should be installed a minimum of 2 ft. o. c.

Wire mesh is the best reinforcement option—Next we place the reinforcement. We use 6-in. by 6-in. wire mesh throughout the slab as reinforcement. Wire mesh is not designed to prevent concrete from cracking. However, it does prevent widening or horizontal separation of cracks that form.

There are many claims that cracking can be eliminated by mixing nylon or steel fibers with the concrete. I've used fiber mesh on a couple of large parking lots, and I have mixed feelings about the long-term results. I still think that wire-mesh reinforcement installed correctly is best.

In most cases wire mesh is pulled up into the wet concrete while it's being poured, ideally 1 in. to 1½ in. above the bottom of the driveway slab. We have the concrete truck place a bead of concrete at a uniform height that supports the wire mesh while the rest of the concrete is being poured (photo center right, facing page). Another method that I prefer (although it's probably more time-consuming) is setting the wire mesh on 1½-in. chairs that keep the wire at a more consistent height.

When the wire mesh is in place, I make sure everything has been prepared properly and that

Working the edges. A crew member with a flanged edging tool works along the perimeter of the slab to consolidate material and to rough-cut the rounded edge.

A bull float smooths the concrete and fills voids. As the crew member in the foreground pulls the wide blade of the bull float over the concrete, surface tension is created to bring water to the surface and fill in imperfections from screeding.

every tool needed for finishing the slab is on hand. The next step is ordering the concrete.

Order the right concrete mix—Concrete mixes vary depending on the application. But for most driveways, we use a six-bag limestone mix (4,000 psi) with approximately 6% air entrainment. Air entrainment is the incorporation of microscopic air bubbles throughout concrete to help prevent scaling, the flaking or peeling that occurs on cured-concrete surfaces.

The mix order should also include the slump requirement, or the wetness of the concrete when it's delivered. Slump is measured on a scale of 1 to 12 with 1 being the driest mix. For most driveways, we request a slump of 4 to 5, which is easy to spread but can be worked shortly after it is poured. After a mix has been prepared to specifications, adding water can weaken it. The concrete supplier is responsible for the slump as well as the strength of the mix, and the concrete should arrive as ordered. If concrete arrives too wet, it can be sent back.

Screeding creates the level of the slab—Trucks in our area are able to distribute concrete pretty evenly by controlling the flow of material and the direction of the chute. We work the concrete along the edges to consolidate it and to remove any voids. Then the concrete is raked to a rough elevation just slightly higher than the forms and expansion joints. We make sure we never get too far ahead of the screeding process so that any excess can be easily raked down to areas waiting for concrete.

Screeding cuts in the grade of the slab and consolidates the concrete before bull floating (photo bottom left, p. 138). It's usually done with a long, straight 2x4 (we sometimes use an aluminum box beam screed rail), slightly longer than the width of the driveway. The 2x4 or screed rail that rides on the forms is pulled across the wet concrete with a side-to-side reciprocating motion.

After two or three yards have been placed, the edges should be hand-floated and cut in with an edging tool (photo right, p. 139). The screeded concrete can now be bull-floated (photo left, p. 139). A bull float is a wide, flat metal float mounted on the end of a long handle. As the float is pushed and pulled over the screeded concrete, the leading edge must be elevated to keep from digging in. If the float is mounted on the handle at a fixed angle, bull-floating can be a real workout. The best bull floats have a blade that rotates back and forth by simply twisting the handle; this design allows the operator to keep the handle at a constant, comfortable angle. As the bull float rides over the concrete, it creates surface tension that brings water to the top, which smooths the slab and fills in minor voids at the same time.

After bull-floating, the concrete should be left alone until all bleed water on top of the wet concrete has evaporated. At this point the concrete should be strong enough to support a crew member on kneeboards, which I'll describe later, and is ready for finishing. Finishing the concrete too early can trap water and create a weak surface with a high water/cement ratio.

Contraction joints give the slab a place to crack—The first part of the finishing process is laying out and cutting the contraction or control joints (photo bottom left). Contraction joints act as score marks in the concrete; they create weak points and encourage any cracks that might develop to occur at the joints. To understand how contraction joints work, I need to explain why concrete cracks.

Concrete begins to crack before receiving any loads whatsoever. As the concrete cures and dries, water is absorbed into base materials and evaporates through the surface, which causes the concrete to shrink or contract. Cracks form in the concrete as a result. Contraction joints provide the relief needed so that these cracks form along a joint instead of randomly in the surface of the slab.

Maximum spacing of contraction joints should follow this rule of thumb. Multiply the thickness of the slab by 2½, and that number represents the maximum distance in feet between joints in any direction. The slab for this driveway was 4 in. thick, so the maximum distance between the joints is 10 ft. (4 x 2.5 = 10).

The depth of the joint should be no less than one-quarter of the thickness of the slab and should be cut in either during the finishing process or immediately afterward. For this driveway we cut the joints with special tools called groovers. We begin by stretching a string between our layout lines and snapping it to leave an impression in the wet concrete. We work the groovers along straightedges to cut in the joints. Thicker slabs require deeper joints that are cut

A special tool cuts the contraction joints. Contraction joints score the slab so that cracking from the curing process occurs at these points rather than at random.

with saws equipped with blades designed to handle fresh, or green, concrete.

Kneeboards distribute body weight when you're finishing fresh concrete—After bull-floating, all of the steps in finishing concrete, including cutting in the contraction joints, require a crew member to work in the middle of the slab on top of the uncured concrete (photo below). To keep from sinking into the fresh concrete, this crew member works on a pair of kneeboards. We make our kneeboards out of ½-in. plywood about 24 in. long and 16 in. wide. We cut the corners off to keep them from digging into the fresh concrete, and we put a handle on one end that helps a lot when the crew member is moving them around on the slab.

After the contraction joints have been cut, the crew works the surface of the slab one last time. For most driveways one crew member works the center of the slab while two work the edges. First they go over the surface with an aluminum or magnesium hand float with a thick blade slightly round in section. This process, known as magging, releases air that might be trapped in the concrete from bull-floating and leaves a smooth, open texture on the uncured concrete.

As a final step, a trowel with a thin, broad metal blade is passed over the surface in large circular strokes. Troweling should leave the surface of the slab smooth and flat. If any slurry from the magging gets into the contraction joints, it will be necessary to go back over them with the groover and blend the edges of the groove into the rest of the slab with a trowel.

A broom gives the driveway a rough surface—We give most of our exterior flat work, including driveways and sidewalks, a broom finish for traction. Right after troweling, a crew member drags a wide broom over the slab in smooth, parallel strokes (photo right). We use a fine-bristled broom made either of nylon or of horsehair. A coarse broom will dig into the surface too deeply and dislodge the aggregate. Because this driveway was double wide, the crew member started at the middle and dragged the broom to the outer edge for each side of the slab. The broom should be cleaned by dipping it in water after each stroke to keep excess concrete from building up in the bristles and changing the texture.

Right after we finish, we spray on Kure-N-Seal (made by Sonneborn; see address p. 139), a combination curing and sealing compound. Used to prevent water from evaporating too rapidly, curing compounds form a membrane on the surface of a slab. Application of curing compound effectively slows the curing rate, and the longer concrete takes to cure, the stronger it becomes. Concrete treated with curing compound also has better resistance to scaling. Sealing compounds prohibit moisture from getting into the concrete once the concrete has cured. Because the concrete's curing takes several days, we recommend not allowing vehicular traffic on the slab for a week. □

Rocky R. Geans is the owner of L. L. Geans & Sons, concrete contractors of Mishawaka, Indiana. Photos by Roe A. Osborn.

A broom finish provides traction. A broom finish is applied to the still-wet concrete after it is troweled. A wide, fine-bristled broom is dragged slowly in parallel strokes from the middle of the slab out and is cleaned with water between strokes.

Kneeboards keep the middle man from sinking in the concrete. The crew works the surface for the last time, going over it first with magnesium floats and then doing a final smoothing with flat-bladed trowels. The crew member in the middle works on plywood boards that distribute his weight and keep him from sinking into the slab.

INDEX

The articles in this book originally appeared in *Fine Homebuilding* magazine. The date of first publication, issue number and page numbers for each article are given at right.